62nd International Scientific Conference on Experimental Stress Analysis (EAN 2024)

AA005215

Full Proceedings

Bozi Dar, Czech Republic
4-6 June 2024

ISBN: 979-8-3313-1442-2

Printed from e-media with permission by:

Curran Associates, Inc.
57 Morehouse Lane
Red Hook, NY 12571

Some format issues inherent in the e-media version may also appear in this print version.

Copyright© (2024) by Czech Society for Mechanics (CSM)
All rights reserved.

Printed with permission by Curran Associates, Inc. (2025)

For permission requests, please contact Czech Society for Mechanics (CSM)
at the address below.

Czech Society for Mechanics (CSM)
Dolejskova 5
Praha 8
182 00
Czech Republic

Phone: +420 266 053 045
Fax: +420 286 587 784

csm@csm.cz

Additional copies of this publication are available from:

Curran Associates, Inc.
57 Morehouse Lane
Red Hook, NY 12571 USA
Phone: 845-758-0400
Fax: 845-758-2633
Email: curran@proceedings.com
Web: www.proceedings.com

EXPERIMENTAL STRESS ANALYSIS 2024

62nd International Conference
June 4th – 6th, 2024, Boží Dar, Czech Republic

EAN 2024
62nd conference on experimental stress analysis

BOOK OF EXTENDED ABSTRACTS

EDITED BY

PAVEL POLACH
LUKÁŠ STUNA

Výzkumný a zkušební ústav Plzeň s.r.o.
(Research and Testing Institute Plzen)

Czech Society for Mechanics
expert group for Experimental Mechanics

The editors did their best to ensure the printing quality of the proceedings. However, some manuscripts did not comply with the proceeding's instructions. Therefore, the editors cannot accept responsibility for the final appearance of these manuscripts.

Authors: Collective Authorship

Published by:
Západočeská univerzita v Plzni / University of West Bohemia in Pilsen
Univerzitní 8, 301 00 Plzeň, Czech Republic

Printed by:
BÍLÝ SLON - COPYDIGITAL s.r.o.
Škroupova 1017/11, 301 00 Plzeň

Copyright © 2024 Výzkumný a zkušební ústav Plzeň s.r.o.

First edition, 2024

© Západočeská univerzita v Plzni / University of West Bohemia in Pilsen

This page intentionally left blank.

Main Subject Areas

- Development of experimental methods in mechanics
- Development of experimental methods in biomechanics
- New methods and applications of deformation and stress analysis in mechanical structures
- New methods and applications of deformation and stress analysis in civil and other structures
- Experiment as a tool for the verification of analytical and numerical methods
- Experimental research and prediction of strength, life, and operational reliability of structures
- Monitoring of operational loadings and operational states of structures and devices
- Mechanics of advanced materials, composites and 3D printing materials
- Methods and means of teaching experimental methods

Program Committee

Members

- Karel Doubrava – Czech Technical University in Prague, CZE
- Peter Frankovský – Technical University of Košice, SVK
- Nikolaj Ganev – Czech Technical University in Prague, CZE
- Radim Halama – Technical University of Ostrava, CZE
- František Holešovský – University of West Bohemia, CZE
- Stanislav Holý – Czech Technical University in Prague, CZE
- Lubomír Houfek – Brno University of Technology, CZE
- Miroslav Hrabovský – Palacký University in Olomouc, CZE
- Róbert Huňady – Technical University of Košice, SVK
- Josef Jíra – Czech Technical University in Prague, CZE
- Kamil Kolařík – Czech Technical University in Prague, CZE
- Aleš Lufinka – Technical University of Liberec, CZE
- Tomáš Návrat – Brno University of Technology, CZE
- Pavel Padevět – Czech Technical University in Prague, CZE
- Iva Petríková – Technical University of Liberec, CZE
- Roman Růžek – Czech Aerospace Research Centre, CZE
- Milan Růžička – Czech Technical University in Prague, CZE
- Jaroslav Václavík – Research and Testing Institute Plzen, CZE
- Robert Zemčík – University of West Bohemia, CZE

Organizing Committee

Members

- Lenka Lopatková
- Radka Musilová
- Pavel Polach
- Jaroslav Václavík

Conference Partners

DEWESoft®

Gantner instruments

HBM HBP měřicí technika

KESACON
- official Kyowa representative -

LABTECH

MICRO-EPSILON

SHIMADZU
Excellence in Science

YOTTAVOLT

Table of Contents

MAIN SUBJECT AREAS IV

PROGRAM COMMITTEE V

ORGANIZING COMMITTEE V

CONFERENCE PARTNERS VI

TABLE OF CONTENTS VIII

X-RAY COMPUTED TOMOGRAPHY OF THE OSCILLATING BEAM 1

P. BENEŠ, T. FÍLA, P. KOUDELKA, D. KYTÝŘ, M. MACHÁČEK, V. RADA, P. ZLÁMAL, D. VAVŘÍK 1

BENDING ANALYSIS OF GLASS FIBER REINFORCED COMPOSITE USING DISTRIBUTED FIBER OPTIC SENSING SYSTEM 7

D. BLAHA, Z. PADOVEC, K. DOUBRAVA, R. KOTTNER, C. NOVOTNÝ, N. SCHMIDOVÁ, M. RŮŽIČKA 7

EXPERIMENTAL VERIFICATION OF THE INFLUENCE OF THICKNESS ON 3D PRINTED SAMPLES ON FRACTURE TOUGHNESS PARAMETERS 15

P. BOČAN, A. JÍRA 15

MODELING THE COMPRESSION BEHAVIOR OF ADDITIVELY MANUFACTURED LATTICE STRUCTURE AND ENERGY ABSORPTION 20

A. BOUALLEG, J. FIORI, D. CIRKL 20

VARIOUS APPLICATIONS OF THE DIGITAL IMAGE CORRELATION METHOD IN THE EXPERIMENTAL ANALYSIS OF BIOMECHANICAL PROBLEMS 25

G. CABAJ, P. BOHÁČ, T. NÁVRAT, P. VOSYNEK, V. APOSTOLOPOULOS, L. KUNZ, J. BENEŠ 25

FLEXURAL BEHAVIOUR OF A GLASS-FIBRE REINFORCED POLYMER COMPOSITE: EXPERIMENT AND SIMULATION 34

J. EZENWANKWO, I. PETRIKOVA, J. ZAK 34

IMPACT OF DIRECT AGEING ON MECHANICAL PROPERTIES AND RESIDUAL STRESSES OF MARAGING 300 STEEL MANUFACTURED BY SLM 40

R. HALAMA, M. KOŘÍNEK, J. ČAPEK, K. TROJAN, J. HAJNYŠ 40

METHODOLOGY AND DETERMINATION OF PARAMETERS FOR MODELING BRITTLE ADHESIVE 45

L. HORÁK, J. KRYSTEK 45

DETERMINATION OF THE DYNAMIC BENDING AND TORSIONAL STIFFNESS OF A BEAM USING ENHANCED FRF 53

R. HUŇADY, P. PALIČKA, E. DUBŇANSKÁ, M. HANČINOVÁ 53

EXPERIMENTAL STUDY OF MECHANICAL PROPERTIES OF TITANIUM ALLOY WITH BETA STRUCTURE 59

A. JÍRA, Z. TOLDE, J. FOJT 59

INFLUENCE OF TEMPERATURE CHANGE ON YOUNG MODULUS OF ELASTICITY OF CFRP 64

B. KROPÍK, M. MATUŠŮ, Z. PADOVEC, T. MAREŠ. 64

FATIGUE LIFE PREDICTION OF 42CRMO4+QT STEEL USING SELF-HEATING EFFECT 69

M. MATUŠŮ, J. PAPUGA 69

ON A POSSIBILITY OF HIGH-RESOLUTION NEUTRON DIFFRACTION OBSERVATION OF TIAL SUBSTRUCTURE 81

P. MIKULA, A. MICHALCOVÁ, V. RYUKHTIN 81

SHRINKAGE OF RAMMED UNFIRED EARTH IN LONG TERM OBSERVATION 86

P. PADEVĚT, B. MUŽÍKOVÁ. 86

ANALYSIS OF THE STRAIN-RATE DEPENDENCE ON THE MECHANICAL PROPERTIES OF POLYPROPYLENE.. 91

P. PALIČKA, R. HUŇADY, M. HANČINOVÁ, E. DUBŇANSKÁ........91

DETERMINATION OF SHEAR PROPERTIES OF 3D PRINTED COMPOSITE 97

P. PAVLÍČEK, M. FUSEK, M. ŠOFER 97

CHARACTERIZATION OF THE INTERFACIAL TRANSITION ZONE BETWEEN THE CEMENT MATRIX AND GLASS POWDER PARTICLES 105

Z. PROŠEK, A. PALIČKA, P. TESÁREK 105

BENDING ANALYSIS OF GFRP COMPOSITE PLATE WITH A HOLE USING CARBON FIBER SENSORS 109

N. SCHMIDOVÁ, K. DOUBRAVA, Z. PADOVEC, D. BLAHA, R. KOTTNER, C. NOVOTNÝ, M. RŮŽIČKA........ 109

VALIDATION OF SIMULATION OF THE TEMPERATURE FIELD OF A PART DURING SELECTIVE LASER MELTING PROCEDURE 116

R. SKÁCELÍK, M. ŠPANIEL. 116

METHODOLOGIES FOR MODELING AND TESTING BUS HYBRID STRUCTURE ROLL-OVER DYNAMICS........ 121

S. ŠPIRK, T. KALINA, J. VÁCLAVÍK 121

BIAXIAL CRUCIFORM TESTING OF FIBER COMPOSITE MATERIALS 129

V. VOMÁČKO, M. ŠULC, J. KOLÁŘ. 129

EFFICIENCY OF REINFORCEMENT MATERIALS ON BUS FRAME STABILITY 136

V. YANOVYCH, J. VÁCLAVÍK.. 136

EAN 2024 62th annual conference on experimental stress analysis

X-ray Computed Tomography of the Oscillating Beam

P. Beneš [1,a], V. Rada [1], M. Macháček [1], P. Koudelka [1], P. Zlámal [1], T. Fíla [1], J. Šleichrt [1], D. Vavřík [1]

[1] *Institute of Theoretical and Applied Mechanics, Czech Academy of Sciences, Prosecká 76, 190 00, Prague 9, Czech Republic*

[a] *benes@itam.cas.cz*

Abstract: This work presents X-ray computed tomography of a harmonically oscillating elastic beam. The tomographic imaging is achieved by obtaining projections of the object in its maximum amplitude to capture the eigenmode of the beam using a triggered computed tomography method synchronized with the oscillation of the beam. Related instrumentation, experimental methodology and data post-processing are presented. Results are demonstrated on the reconstructed 3D images of the polymeric oscillating beam in its deformed shape corresponding with the harmonic excitation using the first eigenfrequency. A comparison with conventional non-synchronized tomography of the oscillating beam in its first eigenfrequency is presented to highlight the fundamental advantages of such tomographical imaging.

Keywords: in-situ computed tomography; high-speed X-ray imaging; modal analysis; eigenmode visualization.

1 Introduction

The X-ray computed tomography (CT) has become a common nondestructive inspection method in many scientific fields. Its 4D CT variant is devoted to monitoring the structural and geometrical changes of the objects during time-dependent processes. In this particular case study, computed tomography is employed as a tool for modal analysis and volumetric visualization of the beam oscillation at its first eigenfrequency.

The standard 4D CT method, which uses conventional laboratory CT scanners, allows for reaching temporal resolution typically in a range of seconds in the best case while the limitation is in most cases caused by CT hardware restrictions. The X-ray source would need to be very intense to effectively reduce the exposure time while the detector would need to have a frame rate in the thousands of frames per second. Here, to perform a tomographical scan of an object, hundreds or even thousands of projections need to be recorded within the temporal resolution time interval. An exception to this temporal resolution restriction is the monitoring of periodically moving objects. In this case, the required number of projections can be repetitively captured within a short time window temporally synchronized with the physical process, thus eliminating the need to capture them all within a single temporal resolution interval.

Several articles related to time-lapse CT measurements of objects subjected to time-dependent processes can be found in the literature. For instance, the 4D CT of a beating heart triggered by an electrocardiogram is discussed in [1]. The combination of motion estimation and motion-compensated reconstruction in 4D-CT imaging has been used to improve CT image reconstruction and reduce X-ray dosage in lungs and livers [2]. The time evolution of beer foam and the muffin baking process have been investigated using CT [3]. Synchrotron-based measurements, such as those discussed in [4], are capable of recording projections very quickly. One of the newest methods is tomoscopy [5], which uses a strong synchrotron source and is capable of taking hundreds of projections per second. The first three mentioned 4D CT methods [1, 2, 3] are suitable for slowly moving or evolving objects. Synchrotron-based methods can achieve higher temporal resolution scans, but such measurement is much more hardware-demanding compared to the use of conventional X-ray tubes.

This study presents a triggered CT (tCT) system that delivers exceptional temporal and spatial resolution of a moving object with the use of a conventional sealed X-ray source. A unique CT data recording protocol where each projection is produced by integrating short individual images captured at the same position of a periodically moving object makes this approach a promising alternative to the established methods. A critical requirement for this approach is a precise and multi-dimensional synchronization of all key elements of the tCT system, including the angular position of the CT scanner's rotary stage and the X-ray detector read-out triggered by the object's actual position. To ensure the precise and reliable acquisition of projections, the system employs real-time hardware synchronization of all components through specialized control software that interfaces with the X-ray detector data acquisition software.

In this work, we present the tCT of a beam oscillating at 4 Hz captured during the amplitude of its periodic motion induced by the exciter. The oscillatory motion is performed by an electromagnetic shaker mounted on the CT rotary stage.

2 Measurement Setup

2.1 Specimen

A slender beam made of epoxy resin PR102 with EM420 hardener with a total mass of 2.79 g was used in the experiments. To achieve optimal material properties and a constant cross-section of the sample curing in a negative mold at a constant temperature of 20 °C and low humidity of approximately 20 % was necessary. Aluminum particles with a diameter of 0.2 - 0.3 mm were added to the resin to improve the beam's contrast in the X-ray projections. Once fixed in the shaker, this beam becomes a cantilever with a length of 188.5 mm from the end of the composite base to the top and constant cross and constant cross-section of 9.9 mm x 1.4 mm. The dimensions of the beam are depicted in Fig 1.

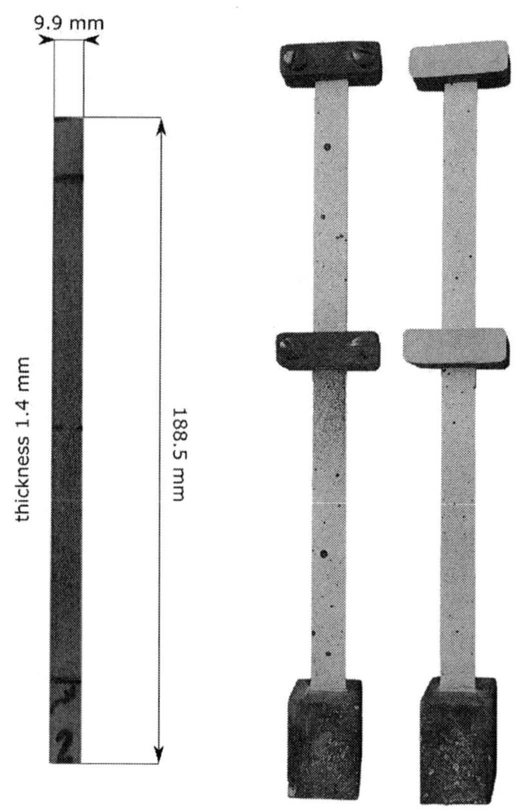

Fig. 1: Sample beam dimensions and front and back view.

This entire beam was bonded into a cubic base made from carbon fiber composite to achieve better repeatability of the measurements through consistency of boundary conditions and the beam length is set to exactly the same value for each measurement.

Due to the display and triggering capabilities and the fact that this is a pilot test, we chose the analogy of a simple analytical model with 2 degrees of freedom. This was obtained by adding two plastic weights to the beam, one in the middle and one at the free end. Both were fixed with plastic screws to avoid potential photon scatter in the reconstructed 3D images. The middle-added weight is a prism with dimensions 17 x 10 x 30 mm, a weight of 5.38 g including the plastic screws, and the center of gravity of the weight was placed 100.5 mm above the end of the cube base. The top added mass was the same shape as the middle mass except the thickness was 8.5 x 10 x 30 mm, the mass including plastic screws was 2.69 g and the center of gravity of the mass was located 183.5 mm above the cube base. The first and second natural frequencies of the cantilever, as measured by laser vibrometry, were 4.15 Hz and 23.75 Hz, respectively.

2.2 tCT Setup

The in-house developed CT system consisting of a Dexela detector and Hamatsu X-ray source was combined with a shaker stage providing harmonic excitation of the specimen. The shaker was mounted on top of the rotary stage and was based on a voice coil actuator (MGV41, Akribis, Singapore) controlled by the Akribis ASD system (Fig. 2). The frequency bandwidth of the shaker can reach up to 50 Hz, maximum stroke is 25 mm, typical position accuracy during harmonic oscillations in position control mode is better than 1.5 μm with position control sensitivity as low as 0.1 µm. The Dexela 1512 NDT CMOS detector with CsI scintillator (Varex Imaging, Germany) was used for the acquisition of the X-ray projections. Its native resolution is 1944 x 1536 pixels with binning–dependent maximum framerate reaching up to 86 Hz. Sealed-type X-ray source Hamamatsu Photonics L10321 with an acceleration voltage range of 40 – 100 kV, maximum target power of 20 W, and focal spot range of 5 μm to 30 μm was used for irradiation of the sample.

Fig. 2: Measurement setup.

A fundamental requirement for a reliable tCT system is precise synchronization of all key components. These include the shaker, the rotary stage, and the detector acquisition trigger and image exposition. The key strategy is to place the image exposure symmetrically around the maximum amplitude of the oscillating beam to minimize negative motion blur effects. As an additional condition, the rotary stage has to be stationary during image exposure. This task was accomplished through the use of an in-house developed control software package working under the LinuxCNC open-source software platform. The tCT system gains better image statistics by taking multiple projections at identical conditions repeatedly and averaging them. A closer description of the tCT method can be found in [6].

3 Tomography of the Beam

The cantilever beam, as described in Section 2.1, was tomographically scanned while oscillating at a first eigenfrequency of 4.15 Hz. The X-ray source for the tCT measurements was operated at 100 kV of acceleration voltage, 200 µA of target current, and a spot size of 30 µm. Detector binning of 2 x 2 was used during the measurements to increase the maximum frame rate and image statistics. The exposure time was set to 20 ms (15 ms is minimal time with 2 x 2 binning, 5 ms added for better image statistics), with averaging performed over 10 projections. The geometry of the CT setup for tCT measurement was as follows: focus to detector distance of 519.4 mm, focus to object distance of 387 mm, resulting geometrical magnification of 1.34, and voxel size of 55.9 µm.in

The tomography scan consisted of 956 projections acquired with an equiangular step size of 0.38 degrees. Following the tCT measurement, a conventional non-triggered CT measurement of a stationary object was performed using the following parameters: an exposure time of 20 ms 1434 equiangular projections without averaging with an angular step size of 0.25 degrees. An overview of all the parameters is summarized in Tab. 1.

Tab. 1: Scanning parameters tCT/non triggered measurements.

Target power [W]	10.7	Acceleration voltage [kV]	100
Spot size [µm]	30	Projections count	956/1434
Binning	2×2	Frame averaging	10/1
X-ray object dist. [mm]	111.95	Magnification	1.314
X-ray detector dist. [mm]	519.23	Voxel size [µm]	113.79

The tCT-triggered measurement was 2h 34min long compared to approximately 24 minutes of the non-triggered measurement. CT reconstruction was performed using a filtered backprojection algorithm (FBP) implemented in the VGSTUDIO MAX software [7].

Due to the length of the investigated beam and the selected geometrical magnification, it was necessary to scan the specimen by half, it resulted in 3D images of its bottom and top parts, that were then numerically merged. In this procedure, the overlapping part of the merged tomographs is approximately estimated by the user and the merging points are selected in the first CT in the overlapping part during the first step. The positions of these points are then determined in the second volume to be fused by digital image correlation. Due to the internal structure of the beam with small metallic particles resulting in possible misidentification of point pairs, the RANSAC method [8] is employed to determine the mutual transformation. After determining the position of the second tomography in the overall resulting volume, the two volumes are combined by linear weighting in the longitudinal direction of the beam as a simplified version of panorama stitching [9] generalized to 3D.

4 Results

Fig. 3 compares reconstructions of the same beam obtained by the standard non-triggered CT method and the presented tCT method. In the standard CT image Fig. 3a, the reconstructed volume represents a superposition of all its deformed states during oscillation randomly distributed between the amplitudes. It can be seen that the object edges are sharper in the amplitude positions as the object statistically spends more time in this position while the internal structure is irrecoverable due to its superposition from the whole oscillatory movement.

The tCT method Fig. 3b, in contrast, offers a clear image of the entire oscillating object at its amplitude positions. This allows for simultaneous visualization of the object's actual oscillation shape and its internal structure.

By enabling simultaneous visualization of the object's shape and internal structure during oscillation, the tCT method presents a significant improvement over the conventional non-triggered CT method. This capability offers valuable insights into the dynamic behavior of oscillating objects, which would be obscured using traditional methods.

Fig. 4 shows the comparison of the vertical section through the beam when CT is measured at a static state, and tCT is measured at the oscillation amplitude of the beam. A distinct material structure is clearly observable in the lower section of the beam. In the upper region, as a result of higher velocity of motion, the structure remains discernible but slightly blurred. Fig. 5 showcases the displacements observed in the amplitude CT scan. These displacements were determined through the computation of differences in the central plane of mass between the amplitude and stationary state CT scans.

a) Standard non-triggered CT

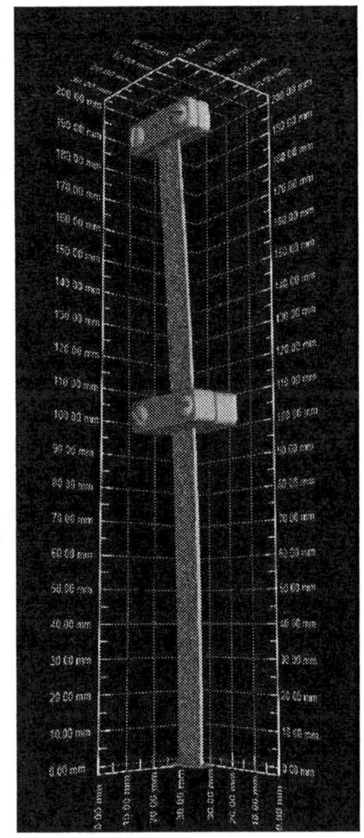

b) tCT of amplitude

Fig. 3: Oscillating beam.

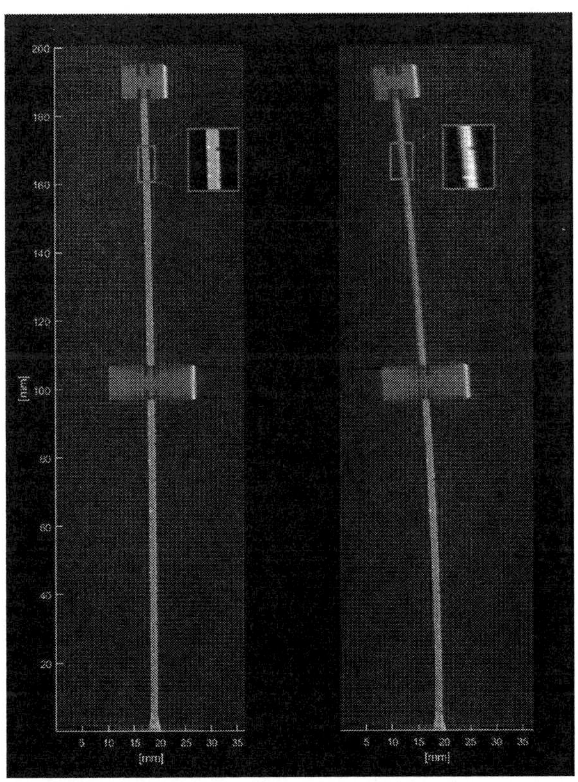

Fig. 4: Slices through CTs from side for static case and amplitude case with material structure detail.

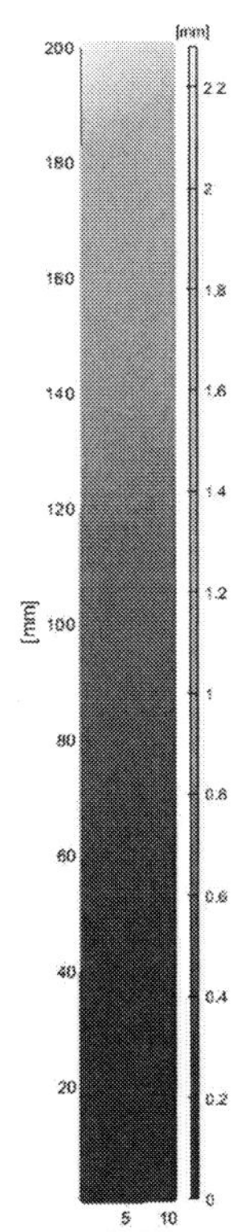

Fig. 5: Displacement values based on difference of the beam's center of mass plane.

5 Conclusion

The principle of the triggered CT (tCT) method was introduced and the method was successfully applied to perform tomography of an oscillating object represented by a slender epoxy resin beam vibrating in its first eigenmode. The measurement of shape in the amplitude was made by a series of 956 projections with an angular step size of 0.38 degrees while each projection was averaged over 10 projections to improve image statistics and thus its quality. The tCT results were compared with a standard CT by analyzing longitudinal sections of the oscillating beam in both its amplitude during vibrations and static state. Both results were comparable while the internal structural details from the conventional stationary tomography scan could be identified in the tCT data of the deformed shape.

The tCT system presented in this study has the potential for various applications, including three-dimensional modal analysis of oscillating objects together with visualization of their internal structure. This approach allows for the identification of defects, such as component separation, that are challenging to distinguish statically. Additionally, it offers the opportunity to validate theoretical dynamic models that are difficult to verify through conventional methods.

In conclusion, the tCT method presents a significant advancement over the conventional non-triggered CT method. It provides a clear and comprehensive image of the oscillating object at its amplitude position. A proof of concept of tCT has been achieved, and with the use of better hardware such as detector Medipix 3 and X-ray tube MetalJet D2+ 160 kV, further work will focus on increasing the temporal resolution and accuracy of the method to obtain relevant modal analysis of more complex objects.

Acknowledgement

This research has been financially supported by the Czech Science Foundation (GAČR) under project No. 22-13811S.

References

[1] M. Gao et al., 4D cardiac Reconstruction Using High Resolution CT Images, Lecture Notes in Computer Science (2011) 153–160, https://doi.org/10.1007/978-3-642-21028-0_19.

[2] Y. Zhang, X. Huang, J. Wang, Advanced 4-dimensional cone-beam computed tomography reconstruction by combining motion estimation, motion-compensated reconstruction, biomechanical modeling and deep learning, Visual computing for industry, biomedicine, and art 2(1) (2019) 23, https://doi.org/10.1186/s42492-019-0033-6.

[3] J. Dewanckele et al., Innovations in laboratory-based dynamic micro-CT to accelerate in situ research, Journal of Microscopy 277(3) (2020) 197–209, https://doi.org/10.1111/jmi.12879.

[4] R. Mokso et al., Following Dynamic Processes by X-ray Tomographic Microscopy with Sub-second Temporal Resolution, AIP conference proceedings, American Institute of Physics, 1365 (1) (2011), https://doi.org/10.1063/1.3625299.

[5] F. García-Moreno et al., Tomoscopy: Time-Resolved Tomography for Dynamic Processes in Materials, Advanced Materials 33(45) (2021), https://doi.org/10.1002/adma.202104659.

[6] D. Vavřík et al., X-ray computed tomography of the periodically moving object, Journal of Instrumentation 19 (2024), http://dx.doi.org/10.1088/1748-0221/19/01/c01045.

[7] https://www.volumegraphics.com/en/products/vgsm.html.

[8] M. A. Fischler, R. C. Bolles, Random Sample Consensus: A Paradigm for Model Fitting with Applications to Image Analysis and Automated Cartography, Communications of the ACM (1981) 381–395, https://doi.org/10.1016/b978-0-08-051581-6.50070-2.

[9] T. Xiang, G.-S. Xia, L. Zhang, Image Stitching with Perspective-preserving Warping, ISPRS Annals of Photogrammetry, Remote Sensing and Spatial Information Sciences, III-3 (2016), 287–294, https://doi.org/10.5194/isprsannals-iii-3-287-2016.

Bending Analysis of Glass Fiber Reinforced Composite Using Distributed Fiber Optic Sensing System

D. Blaha[1,*], Z. Padovec[1], K. Doubrava[1], R. Kottner[2], C. Novotný[1], N. Schmidová[1], M. Růžička[1], P. Balík[1]

[1] *Division of Strength and Elasticity, Fac. of Mech. Engineering, Czech Technical University in Prague; Technická 4, 166 07 Prague, Czech Republic*

[2] *ACO Industries Tábor s.r.o.; Průmyslová 1158, 391 01 Sezimovo Ústí, Česká republika*

* *David.Blaha@cvut.cz*

Abstract: In this study, we conducted a bending analysis of a glass fiber reinforced composite plate with a central hole, using a distributed fiber optic sensing system and other reference methods. The focus was to investigate the change in strain near the hole under bending stress. Our findings provide valuable insights into the behavior of such composites under prescribed load, contributing to their safer and more efficient use in various applications.

Keywords: distributed fiber optic sensing system; strain measurement; glass fiber composite; bending analysis.

1 Introduction

In recent years, water treatment solutions and technologies have been developed to address the growing demand for storing and treating water. The key components of water treatment systems are pressure vessels, which are used to store and treat water. The pressure vessels are typically made of steel, but due to their high weight and susceptibility to corrosion, there is a growing interest in using composite materials for their construction. One of the potential solutions is the use of glass fiber reinforced composited (GFRP) for their high strength to weight ratio. However, the behavior of GFRP under stress is complicated and require detailed analysis to ensure their safe and efficient use especially in critical places such as near holes or other stress concentration points.

A distributed fiber optic sensing system (DFOS) is a powerful tool for measuring strain distribution along the length of a optical fiber. In comparison to conventional measurement methods, DFOS provides dense network of points for strain measurement. In the 1998, Froggatt et al. [5] achieved spatial resolution up to 0.6 cm with tunable laser source to measure strain in the optical fiber. The DFOS system is based on the principle of Rayleigh scattering and is capable of measuring strain with high spatial resolution. The DFOS system is used in various applications, such as structural health monitoring, geotechnical monitoring, and civil engineering [2].

In the Munzke et al. study [3], the authors used a DFOS sensor embedded between layers to monitor the strain distribution in a composite pressure vessel. The monitoring system detected damage 17 000 cycles before the vessel burst. In the Karapanagiotis et al. study [4], the authors successfully embedded a DFOS sensor in a composite pressure vessel to monitor the strain distribution along the length of the vessel including dome and cylinder area.

2 Methods

The measaurement is performed on a composite plate with a central hole. Strain measurements are performed using a distributed fiber optic sensing system (DFOS) and a conventional strain gauges. These methods are used to measure the strain distribution along the hole of the composite plate. The DFOS system is based on the principle of Rayleigh scattering and is capable of measuring strain with high spatial resolution. The strain gauges are used as a reference method. The results from both methods are compared with FEM analysis.

2.1 Sensor installation

The composite plate was made of glass fiber reinforced polyester resin. The sample is 400 mm long, 200 mm wide and 9.76 mm thick. The thickness corresponds to 19 layers of glass fiber. The central hole has a diameter of 80 mm.

The strain gauges were installed to the surface of the composite plate near the central hole in the direction of loading. The DFOS sensor was installed in parallel with the strain gauges, but on the opposite side of the hole. The DFOS sensors were installed in two configurations. The first configuration is a sensor installed on the surface of the composite plate. The second configuration is a sensor installed under the covering glass layer. The sensoric optical fiber was welded to the connector with Fitel S179 fusion splicer and calibrated with Luna software according to the manual [1].

The carbon fiber sensors (CFS) were installed under the covering glass layer. The CFS sensors were installed in the same configuration as the DFOS sensors. The CFS sensors were used to measure the strain. The results from CFS sensors are presented in the Schmidova et al. study [6].

(a) Prepared samples for FO and CFS sensor's installation

(b) Prepared fiber optic and carbon fiber sensors on the plate

Fig. 1: Prepared samples for sensor's installation before covering glass layer.

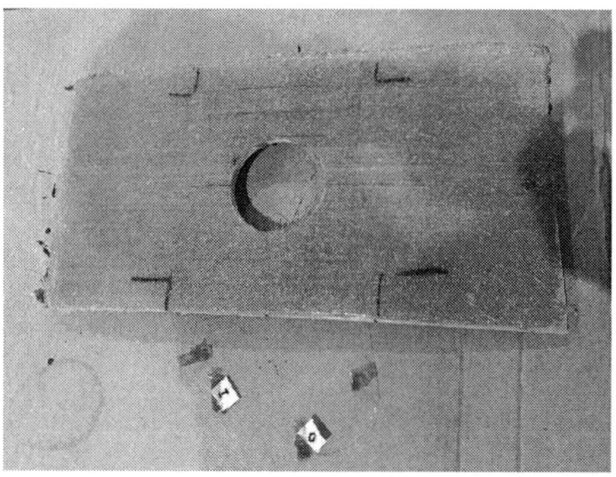

(a) prepared sample with covering glass layer

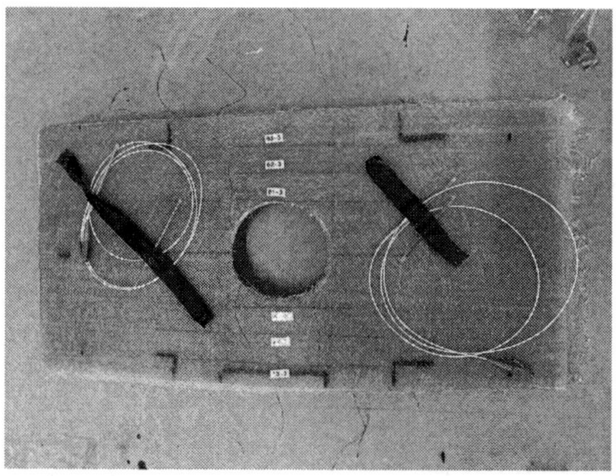

(b) prepared sample with FO connectors

Fig. 2: Prepared samples after sensor's installation.

2.2 Testing

A setup for four-point bending test is following. The distance between the upper supports is 120 mm and the distance between the lower supports is 320 mm. The strain gauges are installed on the upper (compression) side of the plate. The DFOS sensor is installed on both sides of the plate, one on the compression side and one on the tension side.

The testing was subjected to four-point bending on universal testing machine Heckert FPZ100/1. The samples were loaded with a ramp of 1 kN and then unloaded repeatedly. Then the samples were destructively loaded up to 10 kN.

Fig. 3: 4-Point bending test setup.

2.3 FEM analysis

The FEM analysis was performed using the Abaqus software. The composite plate was modeled as a shell element with a thickness of 9.76 mm. The material properties were set to correspond to the glass fiber and polyester resin. The boundary conditions were set to simulate the four-point bending test.

3 Results

Before showing the results, the DFOS data was processed using Python scripts. The data was aranged into separate lines corresponding to the sensor's position on the plate. The bottom side of the plate is loaded in tension and the upper side is loaded in compression. Each figure shows the strain distribution along the sensor's path complet with the sensor's schema. Zero value on the length axis corresponds to the center of the hole. In the figures, there is plotted the mean strain distribution (colored-line) with its standard deviation (colored-area).

Mean deviation was calculated as the percentage difference between the DFOS sensor and the FEM analysis along the sensor's path. The DFOS sensors installed on the surface are named BP4, BP5, BP6, and the DFOS sensors installed under the covering glass layer are named ACO1, ACO2, ACO3A, ACO3B.

The results from the DFOS system show that the strain distribution near the hole is non-uniform. The strain gauges show similar results as the DFOS sensors. The FEM analysis confirms the results from the DFOS system and strain gauges.

In the Figure 4 and Figure 5 are shown the results from all sensors from upper and bottom side in the plane near the edge. It is clear that sensors on the bottom side show better agreement with the FEM analysis. The sensors on the upper side show more variance in the strain distribution. On the compression side (UP), the mean deviation between the DFOS sensor and FEM analysis ranges from 4.2 to 33.9 %. On the tension side (BOTTOM), the mean deviation between the DFOS sensor and FEM analysis ranges from 10.8 to 13.6 %.

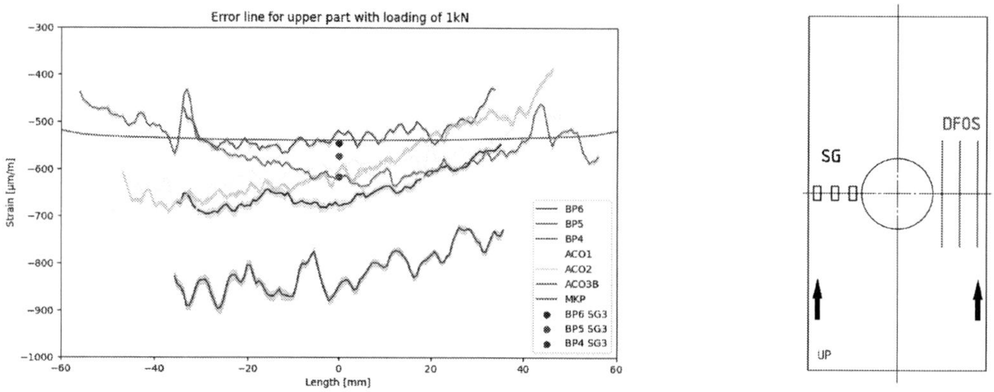

(a) strain results at compression side (b) sensor's schema

Fig. 4: Results from all sensors from upper side in the plane near the edge.

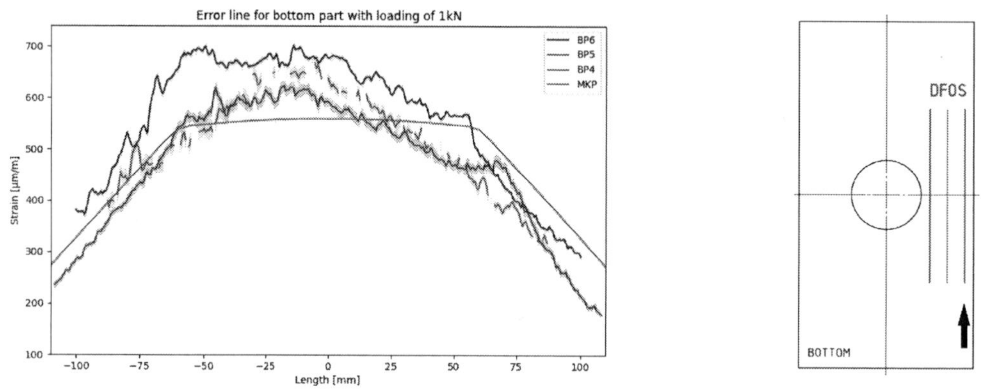

(a) strain results at tension side (b) sensor's schema

Fig. 5: Results from all sensors from bottom side in the plane near the edge.

The results from the sensors in the middle of the plate are shown in the Figure 6 and Figure 7. The mean difference between the DFOS sensor and FEM analysis on the compression side (UP) varies from 8.1 to 30.2 %, see Figure 6. The mean difference on the tension side (BOTTOM) varies from 7.7 to 10.8 %, see Figure 7.

The results from the sensors close to the hole are shown in the Figure 8 and Figure 9. The mean difference between the DFOS sensor and FEM analysis on the compression side (UP) varies from 15.2 to 25.6 %, see Figure 8. The mean difference on the tension side (BOTTOM) varies from 8.5 to 13.3 %, see Figure 9. The sensor ACO3A and ACO3B are excluded from the mean difference calculation due to its closer position to the hole of the plate, which causes the higher deviation from the FEM analysis near the hole area. The trend in the data corresponds across sensors, but probably due to the uneven surface of the composite plate makes the data less reliable.

Based on assumption that the strain concentration points are near the hole, there were installed additional sensors in the plane directly next to the hole. The results are shown in the Figure 10. It is clear that the sensors next to the hole show higher strain values in compare to sensors close to the hole. Even a distance of 10 mm between the sensors near the hole causes a difference of 27 % in the most critical location.

The strain results loaded up to 10 kN are shown in the Figure 11 and Figure 12. The loading up to 10 kN did not cause visible damage to the composite plate. After unloading, only a residual strain of about 100 microstrains was observed in the place around the hole after unloading. The results from the DFOS system are in good agreement with the results from the strain gauges and FEM analysis for the tension side of the plate. The results from the compression side of the plate show more variance. The difference on compression side

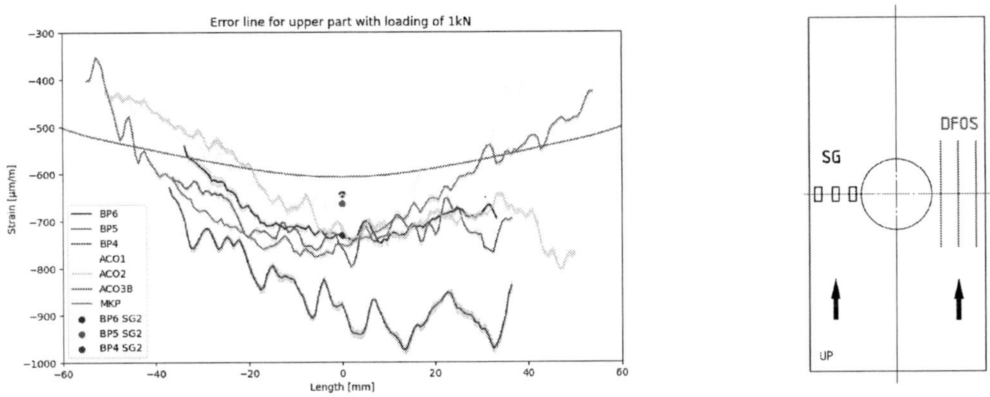

(a) strain results at compression side (b) sensor's schema

Fig. 6: Results from all sensors from upper side in the plane in the middle.

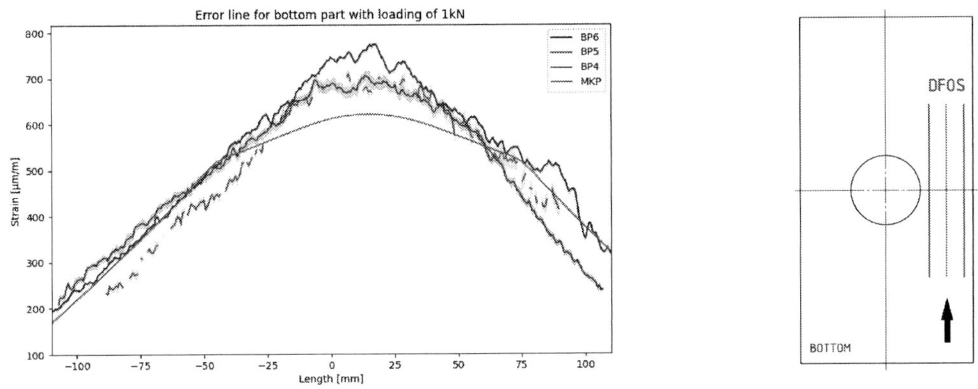

(a) strain results at tension side (b) sensor's schema

Fig. 7: Results from all sensors from bottom side in the plane in the middle.

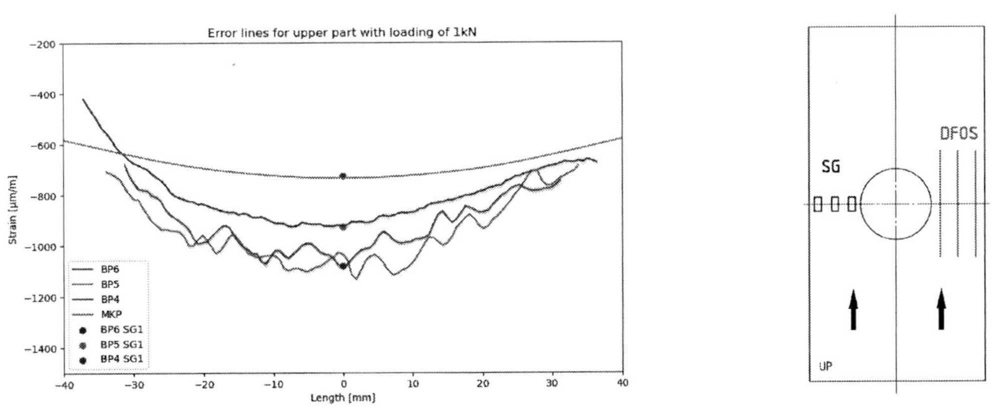

(a) strain results at compression side (b) sensor's schema

Fig. 8: Results from all sensors from upper side in the plane close to the hole.

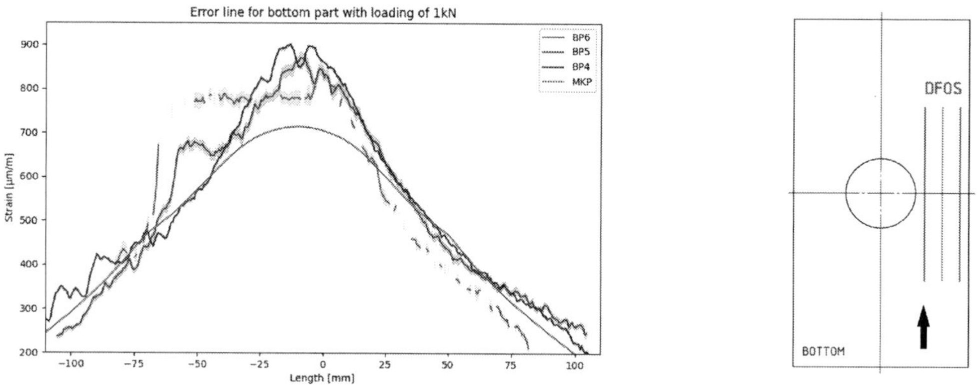

(a) strain results at tension side (b) sensor's schema

Fig. 9: Results from all sensors from bottom side in the plane close to the hole.

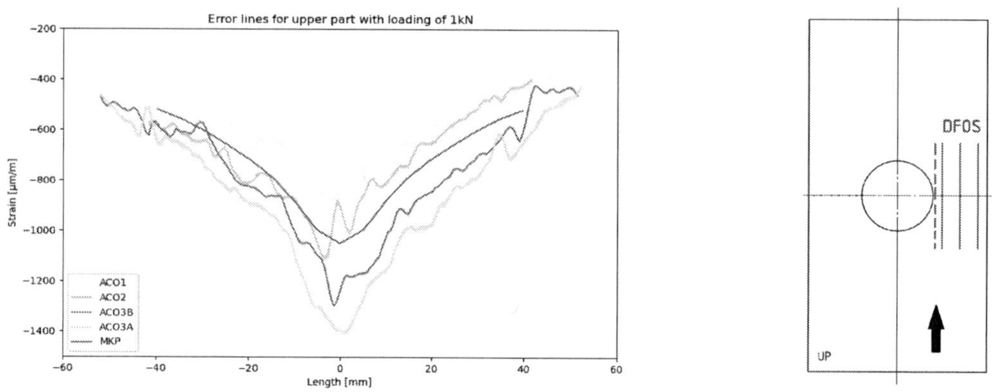

(a) strain results at compression side (b) sensor's schema

Fig. 10: Results from all sensors from upper side in the plane next to the hole.

(UP) between the DFOS sensor and FEM analysis varies from 9.3 to 19.8 %, see Figure 11. Mean difference is 16.4 %. The difference on tension side (BOTTOM) varies from 0 to 14.6 %, and the mean difference is 6.7 %, see Figure 11.

Conclusion

The results of the bending analysis of the GFRP composite plate with a central hole show that the strain distribution near the hole is non-uniform. The results from the DFOS sensors are in good agreement with the results from the strain gauges and FEM analysis on tension side, but show more variance on the compression side. The results from the strain gauges range across the variance of the DFOS sensors.

The sensors installed under the covering glass layer showed better agreement with the FEM analysis than the sensors installed on the surface of the composite plate. This can be attributed to the uneven surface of the composite plate, which makes the data less reliable. This can be reason why the sensors BP4, BP5, BP6 show more variance in the strain distribution than the sensors ACO1, ACO2, ACO3A, ACO3B. It is important to note that the sensors ACO1, ACO2, ACO3A, ACO3B are installed under the covering glass layer only on tension side of the plate, which can cause bigger deviation on compression side.

Loading up to 10 kN did not cause visible damage to the composite plate, although residual strain was observed after unloading.

Applicability of the DFOS system for strain measurement in the composite plate was confirmed, particularly

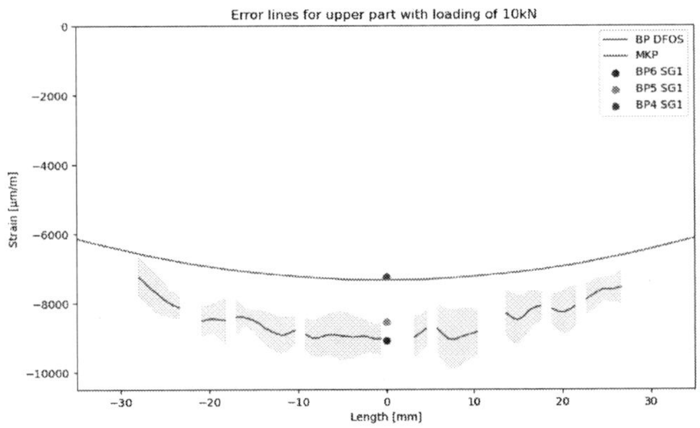

(a) strain results at compression side

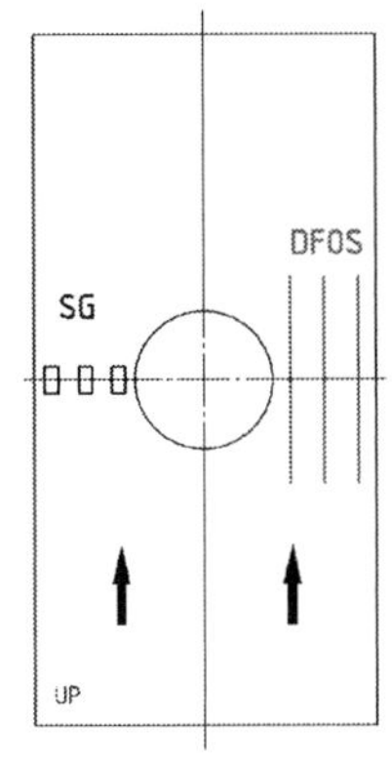

(b) sensor's schema

Fig. 11: Results from all sensors from upper side in the plane close to the hole.

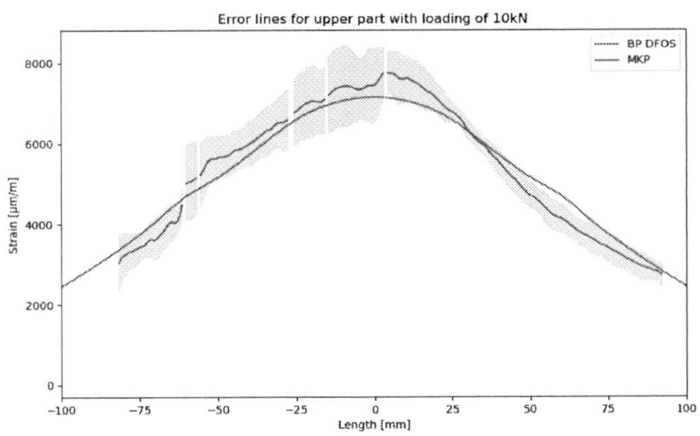

(a) strain results at tension side

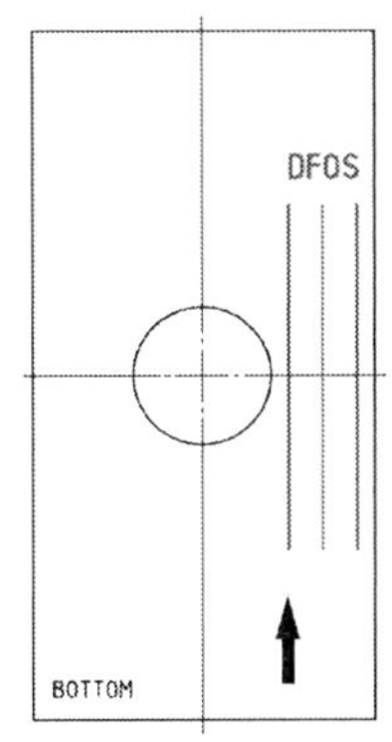

(b) sensor's schema

Fig. 12: Results from all sensors from bottom side in the plane close to the hole.

for the strain concentration points. The method provides valuable insights into the behavior of such composites, contributing to their safer and more efficient use in water treatment applications.

Future work will focus on the development of a connectors and housing for the DFOS and CFS sensors to make them more robust and easier to install. The sensor's ability to measure damage in the composite plate will be tested in the future.

Acknowledgement

The project TN02000010 of National Competence Centre of Mechatronics and Smart Technologies for Mechanical Engineering is co-financed from the state budget by the Technology Agency of the Czech Republic within the National Centres of Competence Programme.

Financial support was provided for this study by the Grant Agency of the Czech Technical University in Prague; grant no. SGS24/123/OHK2/3T/12.

References

[1] Luna Technologies. Optical Backscatter Reflectometer 4600. Technical report, 06 2006.

[2] Boller, C., Chang, F., Fujino, Y. Encyclopedia of structural health monitoring. Wiley-Blackwell, Hoboken, NJ, January 2009.

[3] Munzke, D., Duffner, E., Eisermann, R., Schukar, M., Schoppa, A., et al. Monitoring of type IV composite pressure vessels with multilayer fully integrated optical fiber based distributed strain sensing. Online. Materials Today: Proceedings. 2021, roč. 34, s. 217-223. ISSN 22147853. Dostupné z: https://doi.org/10.1016/j.matpr.2020.02.872.

[4] Karapanagiotis, C., Schukar, M. and Krebber, K. Distributed fiber optic sensors for structural health monitoring of composite pressure vessels. tm - Technisches Messen, 2024, Vol. 91 (Issue 3-4), pp. 168-179. https://doi.org/10.1515/teme-2023-0170

[5] Mark Froggatt and Jason Moore. High-spatial-resolution distributed strain measurement in optical fiber with rayleigh scatter. Appl. Opt., 37(10):1735–1740, Apr 1998.

[6] Schmidová, N., Doubrava, K., Padovec, Z., Blaha, D., Kottner, R., et al. Bending Analysis of GFRP Composite Plate with a Hole Using Carbon Fiber Sensors. In: EAN 2024 62tn annual conference on experimental stress analysis - Conference Proceedings. 2024.

Experimental verification of the influence of thickness on 3D printed samples on fracture toughness parameters

P. Bočan[1,*], A. Jíra[1]

[1] *Department of Mechanics, Faculty of Civil Engineering, Czech Technical University in Prague, 166 29 Prague, Czech Republic*

* *corresponding: petr.bocan@fsv.cvut.cz*

Abstract: Currently, additive manufacturing technologies are widely used in biomedicinal applications. 3D printing technology brings great benefits to this industry as it can produce complex and diverse shapes. The problem then arises with very thin samples, where this technology is still not perfect and shows relatively large uncertainty and manufacturing defects. The main aspect of this research is to test the fracture toughness of very thin polyamide specimens fabricated by 3D printing using SLS technology. The geometry, experimental procedure and evaluation of the results were designed according to EN ISO 12737. The experiment took into account the orthotropy of the print by using different printing directions. Then the resulting data from the experiments were compared with the data from the numerical analysis.

Keywords: fracture mechanics; 3D printing; polyamide PA12; digital image correlation; numerical analysis

1 Introduction

In today's era of rapidly advancing technology and innovation, 3D printing has an indispensable place in manufacturing and scientific research across engineering disciplines. Material engineering and biomedicinal application are one of the key areas where this technology is being used. Titanium alloy 3D printing technology enables to creation of any shape of future implant, the design of an optimized surface and the fast individual manufacture of a specific replacement for a specific patient.

For the purposes of implants and their surfaces, it is necessary to create very thin structures with thicknesses ranging from 150 to 600 μm. This pushes the limits of current 3D printing capabilities, and the resulting mechanical properties do not align with theoretical expectations [1]. Current research often focuses on conventional mechanical parameters, but for obtaining a broader range of characteristics, it is necessary to expand experimental efforts [2, 3, 4]. This expansion is motivated by the subsequent development of a material model that corresponds to the reality of 3D printing. In clinical practice, metallic materials are primarily used for implants but are expensive to produce, and as this research is a pilot study it will investigate polymer samples (PA12 material) produced by 3D printing SLS technology. One of the main motivation for this research is to test fracture toughness, which is a key parameter for assessing the material resistance to accidental loading and crack propagation. The research addresses not only the deep significance of fracture toughness methodologies, but also the design and optimization of methods and procedures for testing, evaluation and applying this knowledge to metallic specimens.

2 Fracture toughness results for PA12 material

Fracture mechanic is the study of engineering mechanical where the resistance of a material to crack propagation is measured and is called fracture toughness. Fracture toughness values can serve as one the main material properties and quality for typical engineering structure [5]. Fracturing of material is cased by the finite strength of atomic bonds. When a material begins to fail, cracks are responsible for amplifying local stress states that lead to fracture [6].

In fracture mechanics studies, two main parameters called fracture energy G and stress intensity factor K are often determined and presented [7]. The fracture energy is defined as the amount of energy needed to create one unit area of a continuous crack and is the fundamental quantity that governs crack propagation [8]. While stress intensity factor is based on three different modes (mode I, mode II, mode III) and can specify the mechanical behaviour of material with crack [9].

2.1 Geometry and pre-test

The geometry of the sample and the calculation of the fracture toughness values was designed according to EN ISO 12737 Metallic materials - Determination of plane-strain fracture toughness. The thickness condition according to the standard was not included in the geometry because investigation of fracture toughness of very thin specimens. The calculation of the fracture toughness value K_Q is given in the Equation 1.

$$K_Q = \frac{F_Q}{B\sqrt{W}} * f(a/W) \tag{1}$$

where F_Q is determined force from the load-distance diagram given in kilonewtons, B is specimen thickness and W is length of a ligament, both given in centimeters. Last parameter in calculation is geometry factor $f(a/W)$ (Eq. 2) which is dimensionless fuction of a/W.

$$f(a/W) = (2 + a/W) * \frac{0,886 + 4,64(a/W) - 13,32(a/W)^2 + 14,72(a/W)^3 - 5,6(a/W)^4}{(1 - a/W)^{\frac{2}{3}}} \tag{2}$$

The 3D models geometry was created in ARCHICAD 25 software using the Profile Manager and were utilized to create 4 different geometries (31.25x30; 37.5x36; 43.75x42; 50x48 mm) with 6 different thicknesses (0.50; 0.75; 1.00; 1.25; 1.50; 2.00 mm). These 3D models were saved as .stl files and exported to Synterit studio, which created S-code for SLS 3D printing. After printing, the specimens were cleaned and prepared for the pre-test experiment. For the pre-test experiment, 96 specimens were printed and a MARK 10 load press was used for testing. For consistency of results, the crack length a was determined as a constant at this stage of research as $a = 0.5 * W$. For the further experiment, 2 specimen geometry were selected which had the most linear increase in fracture toughness values with varying specimen thickness and were best to handle in experiment. Figure 2 shows a diagram of fracture toughness versus thickness of each specimen geometry on the left and the 2 selected specimens SAMPLE 2 and SAMPLE 3 for the following experiment are shown on the right. Figure 1 show the geometry of selected SAMPLE 2 and SAMPLE 3.

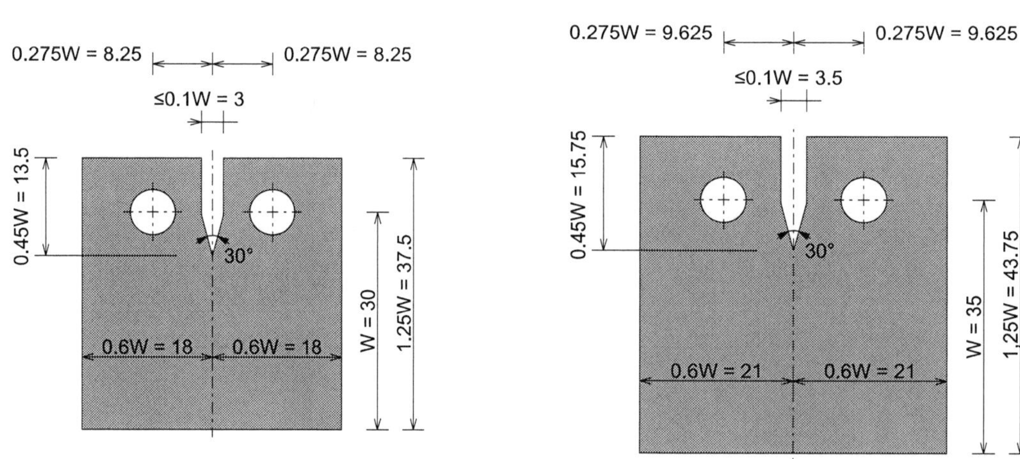

Fig. 1: Example of selected samples geometry according to EN ISO 12 373 designed in ARCHICAD 25 software, SAMPLE 2 on the right, SAMPLE 3 on the left [10].

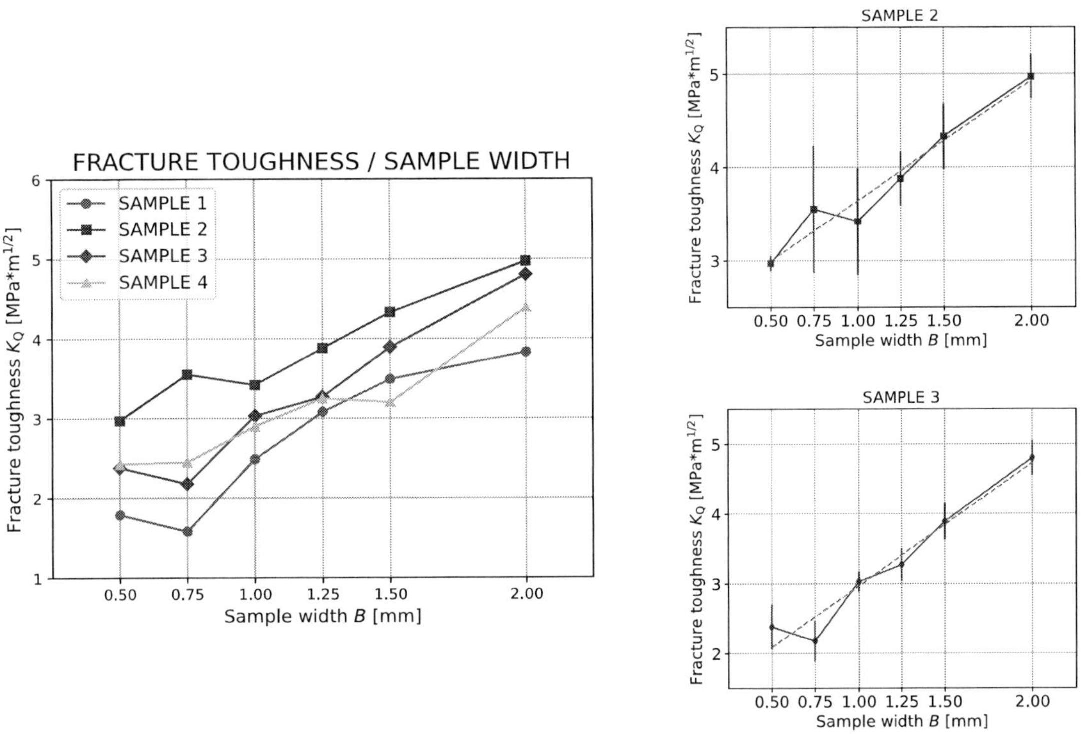

Fig. 2: Fracture toughness results from the pre-test experiment. Summary of fracture toughness values versus width of each specimen geometry (left) and values for selected geometry to the following experiment (right) [10].

2.2 Experimental analysis

In the next step of my research, I proposed to take into account the orientation of printing. The specimens were printed in horizontal (H) and vertical (V) direction which is shown on Figure 3. For both print orientation, 6 pieces were printed for each thickness (0.50 – 2.00 mm), giving a total of 140 printed specimens.

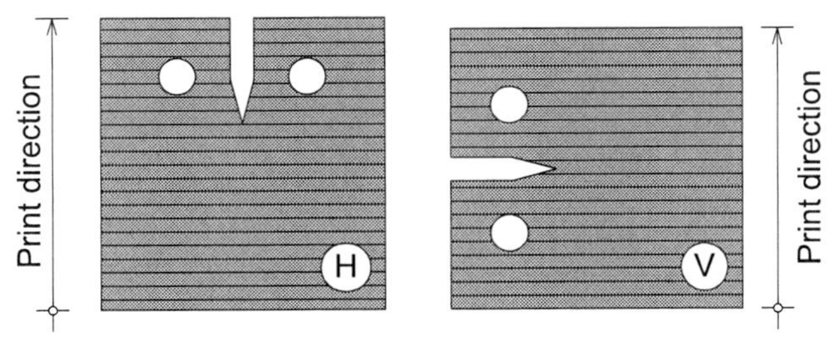

Fig. 3: Demonstration of the two types of different sample printing orientations [10].

This experiment involved a MARK 10 load press where the loading rate was set at 4 mm/s. The experiment was supplemented with 3D DIC, which led to provide the sample with a speckled pattern, where I used combination of white and black color. This pattern is used to mark points from which the displacement of the specimen can be monitored and subsequently the stress and strains values can be calculated [11]. In some cases, especially for thicker specimens, the crack has destroyed the specimen along its entire length. 3D DIC was used to determine the length of the crack a before the specimens were fully ruptured.

2.3 Numerical analysis

For the two specimen geometries (SAMPLE 2, SAMPLE 3), only numerical models for widths of 0.50, 1.25 and 2.00 mm were created. First, I determined the material model of PA12 material. Only the Young's modulus E and the Tension strength f_t are given in the technical datasheet. For the other material properties, I had to do a search of other experiments and determine the values from them, where mainly looked for the values of Poisson's ration μ, Compression strength f_c and Fracture energy G_f.

The numerical model was created in GiD 16.0.6. and the calculation was performed using ATENA studio. A mesh with and element size of 0.25x0.25 mm was used in the most important region below the tip. For all 6 models, the calculation was performed with a linear and quadratic element mesh. Model was loaded with displacement and the Newton-Raphson method was used in calculation.

3 Conclusion

This section summarizes the results form the experiment and numerical analysis. Table 1 shows the summary of fracture toughness values from the experiment calculated according to EN ISO 12 737. The results show a consistent increase for the horizontally printed samples from a certain width and very inconsistent result for thinner widths. The values for vertically printed samples are around 1.00 MPa$\sqrt{\text{m}}$ independent of the specimen width.

Tab. 1: Summary of experiment results of K_Q for PA12 material printed in horizontal and vertical directions with their deviation of results (H - horizontal print orientation, V - horizontal print orientation) [10].

Thickness [mm]	0.50	0.75	1.00	1.25	1.50	2.00
SAMPLE 2 - H	2.0 ± 0.6	2.64 ± 1.12	1.65 ± 0.23	1.87 ± 0.09	2.16 ± 0.45	2.37 ± 0.31
SAMPLE 2 - V	0.99 ± 0.37	0.91 ± 0.38	0.93 ± 0.11	1.35 ± 0.35	0.94 ± 0.07	1.09 ± 0.15
SAMPLE 3 - H	1.54 ± 0.22	1.33 ± 0.48	1.5 ± 0.15	1.73 ± 0.38	2.01 ± 0.33	2.2 ± 0.36
SAMPLE 3 - V	1.2 ± 0.41	0.72 ± 0.11	1.3 ± 0.42	1.03 ± 0.06	1.21 ± 0.15	1.09 ± 0.15

Figure 4 shows a comparison of experimental data and numerical analysis for SAMPLE 2 with widths of 0.50 and 2.00 mm. Agreement between experimental and numerical analysis was not expected. One of the main reasons for discrepancies in results is the imperfect numerical material and geometric model that fails to account for imperfections in 3D printing (such as porosity) and the orthotropic properties of oriented 3D printing. The problem of 3D printing is that the technology is not yet good enough for the individual powder beads on the surface of the sample to melt together sufficiently. These powder beads can then "fall off" when the sample is handled, which weakens the thickness of the sample. This aspect is then fatal for very thin samples.

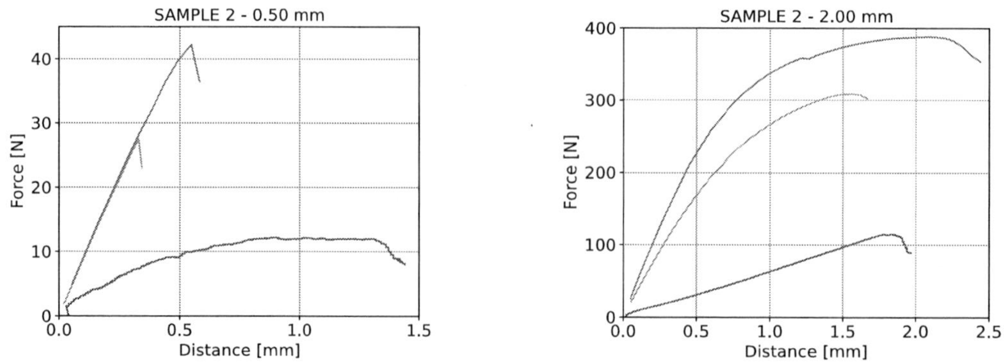

Fig. 4: Load-displacement graphs for one selected sample from the experiment and 2 types numerical models, where the orange line is the numerical model with linear elements, the green line is the numerical model with quadratic elements and the blue line is the experimental data [10].

Acknowledgement

The financial support by Czech Technical University in Prague (SGS project No. SGS23/152/OHK1/3T/11) is gratefully acknowledged.

References

[1] Jíra, A. et al. Mechanical properties of porous structures for dental implants: Experimental study and computational homogenization, Materials, 2021, 14(16). doi: 10.3390/ma14164592

[2] Hájková, P., Jíra, A. Micromechanical analysis of complex structures by nanoindentation, 5th Annual Conference on Nanomaterials and Nanotechnology in Civil Engineering, Prague, 2016, 60-65. doi: 10.4028/www.scientific.net/KEM.731.60

[3] Jíra, Aleš, et al. Mechanical properties of porous structures for dental implants: experimental study and computational homogenization. Materials, 2021, 14.16: 4592. doi: 10.3390/ma14164592

[4] Jíra, Aleš, et al. Experimental investigation of defects of thin 3D-printed plates. In: 60th Annual Conference on Experimental Stress Analysis. Prague, Czech Republic, 2022.

[5] Zhu, Xian-Kui; Joyce, James A. Review of fracture toughness (G, K, J, CTOD, CTOA) testing and standardization. Engineering fracture mechanics, 2012, 85: 1-46. doi: 10.1016/j.engfracmech.2012.02.001

[6] Li, Guodong, et al. Fracture toughness of thermoelectric materials. Materials Science and Engineering: R: Reports, 2021, 144: 100607. doi: 10.1016/j.mser.2021.100607

[7] Soderholm, Karl-Johan. Review of the fracture toughness approach. Dental Materials, 2010, 26.2: e63-e77. doi: 10.1016/j.dental.2009.11.151

[8] de Borst, Rene. Fracture in quasi-brittle materials: a review of continuum damage-based approaches. Engineering fracture mechanics, 2002, 69.2: 95-112. doi: 10.1016/S0013-7944(01)00082-0

[9] Toribio, J., et al. A critical review of stress intensity factor solutions for surface cracks in round bars subjected to tension loading. Engineering Failure Analysis, 2009, 16.3: 794-809. doi: 10.1016/j.engfailanal.2008.06.023

[10] Bočan, Petr. Experimental verification of the influence of thickness on 3D printed samples on fracture toughness parameters. Praha, 2024. Diplomová práce. České vysoké učení technické v Praze, katedra mechaniky. Vedoucí práce doc. Ing. Aleš Jíra, Ph.D.

[11] Zhao, Jianlong; Sang, Yong; Duan, Fuhai. The state of the art of two-dimensional digital image correlation computational method. Engineering reports, 2019, 1.2: e12038. doi: 10.1002/eng2.12038

Modeling the Compression Behavior of Additively Manufactured Lattice Structure and Energy Absorption

A. Boualleg [1,a], J. Fiori [2], D. Cirkl [1]

[1] *Technical University of Liberec, Czech Republic*
[2] *Universidade Estadual Paulista,Brazil*
[a] *abdelmadjid.boualleg@tul.cz*

Abstract: Lattice structures have garnered significant attention due to their lightweight nature and favorable mechanical properties. They find widespread application across various industries and engineering fields, such as automotive, aerospace, marine, and medical. Notably, the deformation of lattice structures allows the dissipation of substantial amounts of energy. In this work, we need to understand the mechanical behavior of the lattice structure. To do that, we need to design the structure in the software and then choose the right flexible material for printing the model thereafter the model is subjected to the experiments. The finite element method is included for comparison of the experiment and simulation.

Keywords: lattice structure; TPU; viscohyperelastic; 3D printing.

1 Introduction

The lattice structures, characterized by their intricate network of interconnected struts, have gained prominence across various engineering domains due to their lightweight nature and unique mechanical properties [1]. Additive manufacturing technologies have enabled the precise fabrication of lattice structures with complex topologies, paving the way for advanced design and production of such structures [2]. These structures often manufactured through advanced techniques like additive manufacturing (AM), present an innovative approach to design challenges. The lattice geometry, with its high surface area and low material volume, offers an excellent compromise between structural integrity and weight efficiency [3]. Lattice structures have been employed in various industrial applications, including scaffolds for tissue and bone replacement, automotive and aerospace components for noise isolation and weight reduction, and a range of protective uses [4]. This work proposes the study of thermoplastic polyurethane (TPU) lattices under loading and unloading cycle compression, along with numerical analysis through FEA. A face-centered cube (FCC) unit cell was selected to design and build the lattice configurations. A commercially available TPU was chosen as the base material due to its flexibility and high support of the load on the structure while avoiding its failure. The tensile test samples of the parent material were manufactured via and later tested under seven cycles of loading, while the lattices were assessed under loading and unloading compression conditions. The mechanical characterization of the base material was used to fit with the Parallal Reological Framework (PRF) to determine the hyperelastic and viscoelastic material constant, which was later used as the input for the numerical simulation of the structure. The experiment and the simulation provided a good agreement. Additionally, the dissipation energy was investigated.

2 Materials and Methods

2.1 Design

The lattice structure sample was designed in Abaqus Software with (10 x10 x .9,6) mm cells with a strut diameter size of 2 mm see Fig. 1(a) and a cube size with a base (60 x 60) and height of 48 mm as presented in

Fig. 1(b). This cube was printed using a 3D printer with selected TPU material. For the material characterization, three dog bone samples were printed and subjected to tensile tests.

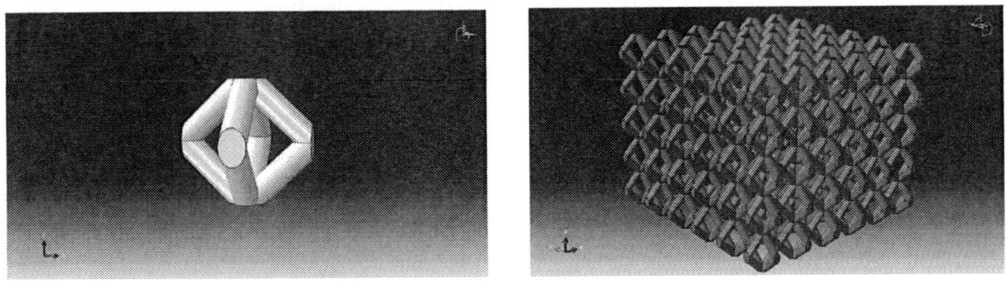

a) One unit cell b) FCC structure

Fig. 1: The design of the lattice structure: (a) unit cell, (b) FCC structure.

2.2 Experiment and Fitting

For printing the lattice structure the TPU material is used. The mechanical properties of the material were performed on the three bone shape samples. The cyclic tension tests were carried out through loading-unloading cycles followed by a final loading phase where the strain was kept constant for a period, allowing the stress relaxation. The cycles were applied at three different strain levels: 0.1 and 0.15, with various strain rates: 0.5 10^{-2} and 0.9 $10^{-2}s^{-1}$. The experiment and fitting results are presented in Figs. 2 and 3.

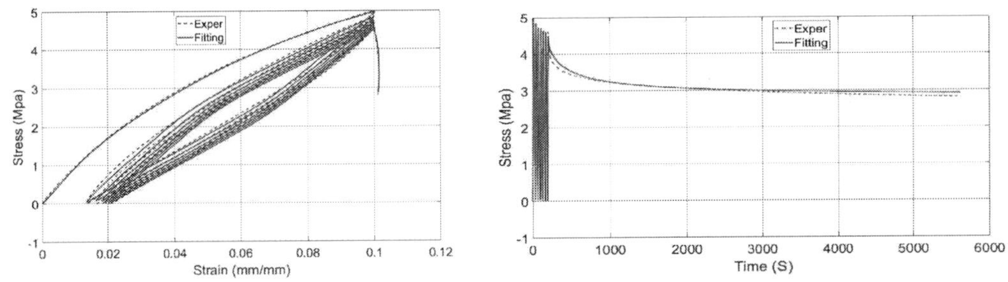

a) Stress-strain curve fitting b) Stress-time curve fitting

Fig. 2: The fitting result at strain 0.1 and strain rate 0.5 10^{-2} s^{-1}.

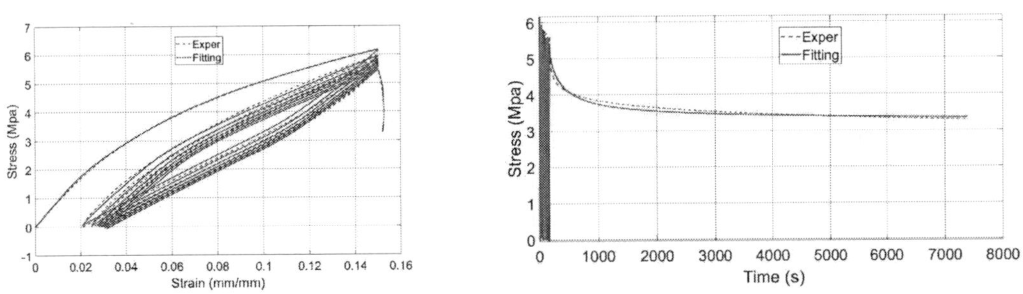

a) Stress-strain curve fitting b) stress-time curve fitting

Fig. 3: The fitting result at strain 0.15 and strain rate 0.9 10^{-2} s^{-1}.

Tab. 1: Hyperelastic parameters (two-terms Mooney).

C1	C2
7.24467	0.34267

Tab. 2: Viscoelastic parameters.

Viscoelastic parameters	Bergstrom-Boyce (Chain B)	Bergstrom-Boyce (Chain C)
nkt	6.38584	9.76011
N	30.2971	21.2684
C_1	6.49168e-05	0.158953
C_2	-0.217285	-9.39227e-06
m	2.18651	3.6405

Tab. 3: Mullins effect model parameters two-terms.

Terms	Scalar factors	Relaxation parameters
1	0.34267	0.189211
2	0.385533	3.68207

3 Results and Discussion

The lower plate compression was constrained in X, Y, and Z axis directions. In contrast, the upper plate was free to move in all directions and allowed to descend along the Z-axis direction of the sample till the desired displacement. The boundary conditions were established in correspondence to the testing conditions used in the experiments. This included matching the step size, time settings, and sample size.

The deformation behavior of different strain levels and stress zones under compression is shown in Fig. 4. For example, in displacement U=4.5 mm it is observed that the higher stress acts on the strut located near the lower and bottom plates, the struts begin to bend and buckle, afterword can be seen when the displacement at U=14.5 the unit cells begin to collapse by elastic buckling and crushing. In the final stage U=24, the lattice cells impinge upon each other.

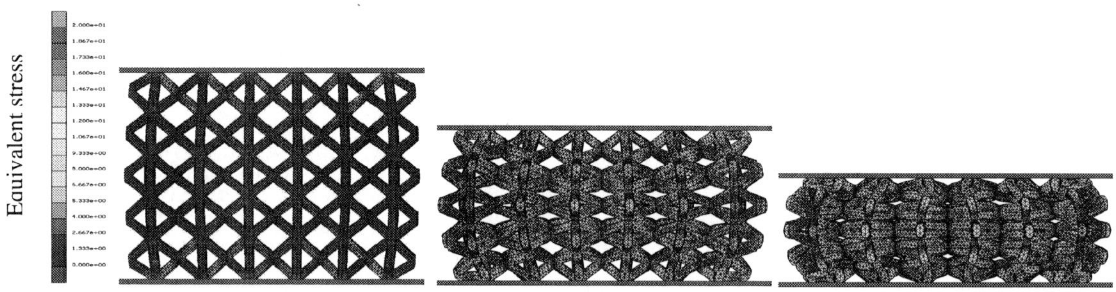

Fig. 4: Compressive deformations of FEM lattice structure at different strain levels.

The cyclic stress-strain curves for a maximum compression displacement of 24 mm were obtained at a strain rate of 0.01 s^{-1}. Figure 5 compares the experimental results with the simulations. Qualitatively, the results are satisfactory, as the FEM simulations captured the experimental behavior well. However, the FEM results are slightly higher than the experimental data.

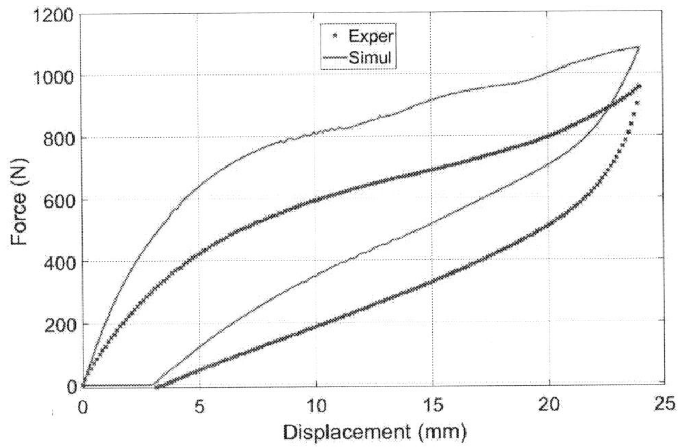

Fig. 5: Comparison of force-displacement the experiments and simulation at strain rate 0.01 s^{-1}.

This discrepancy is likely due to the influence of the printing process on the dog bone tensile sample used for identifying the material parameters. Figure 6 presents the damping force from the experiments, the comparison of damping force between the experiment and the simulation demonstrates a good agreement, as illustrated in Fig. 7. This damping force is caused by the material's viscoelastic properties.

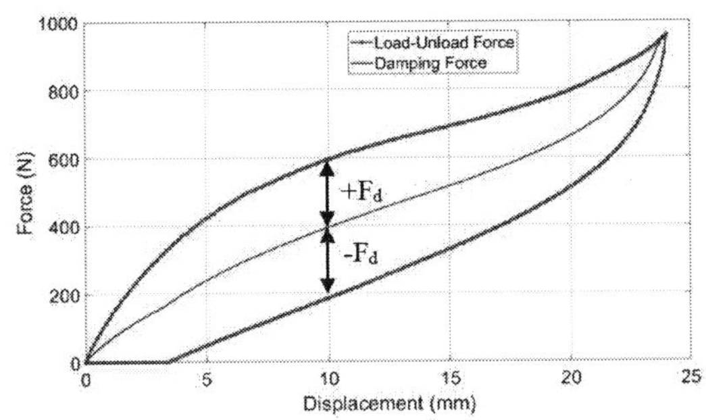

Fig. 6: Presenting the damping force from experiments.

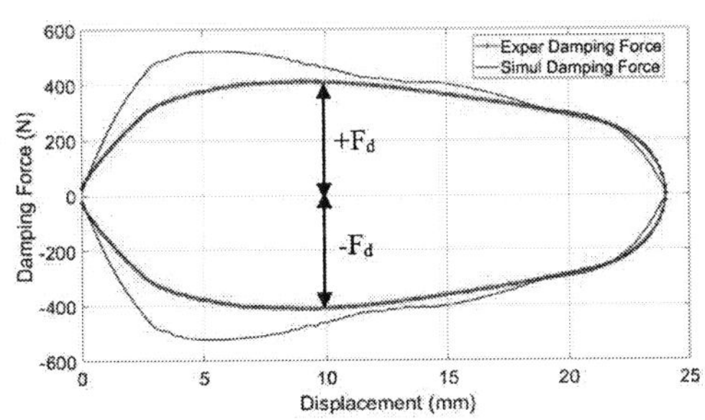

Fig. 7: Compare the damping force of the experiments and simulation.

4 Conclusion

In conclusion, the experiment was conducted on lattice structures under uniaxial compressive force using both experiment and simulation. The TPU material used for printing the samples exhibits characteristics of viscoelasticity with damage. A comparison between the simulation and experimental data revealed a good in force-displacement relationships. Overall, the study highlights the importance of considering material properties and energy absorbed by the structure, which are increasingly being utilized in various engineering applications for their lightweight and efficient design.

Acknowledgment

This publication was written at the Technical University of Liberec as part of the project "Research of advanced materials, and application of machine learning in the area of control and modeling of mechanical systems" nr. SGS-2022-5072 with the support of the Specific University Research Grant, as provided by the Ministry of Education, Youth and Sports of the Czech Republic in the year 2022.

References

[1] J. León-Becerra, O. A. González-Estrada, J. Quiroga, Effect of Relative Density in In-Plane Mechanical Properties of Common 3D-Printed Polylactic Acid Lattice Structures, ACS Omega, 6(44) (2021) 29830–29838, doi: 10.1021/acsomega.1c04295.

[2] P. Wang, F. Yang, J. Zhao, Compression Behaviors and Mechanical Properties of Modified Face-Centered Cubic Lattice Structures under Quasi-Static and High-Speed Loading, Materials 15(5) (2022), doi: 10.3390/ma15051949.

[3] N. Thirunavukkarasu, J. Gao, S. Peng, A. Laroui, L. Wu, Z. Weng, Mechanically robust 3D printed elastomeric lattices inspired by strong and tough hierarchical structures, Additive Manufacturing 66 (2023), doi: 10.1016/j.addma.2023.103451.

[4] J. A. Diosdado, De la Peña et al., Low Impact Velocity Modeling of 3D Printed Spatially Graded Elastomeric Lattices, Polymers 14(21) (2022), doi: 10.3390/polym14214780.

Various Applications of the Digital Image Correlation Method in the Experimental Analysis of Biomechanical Problems

G. Cabaj [1,a], P. Boháč [1], T. Návrat [1], P. Vosynek [1], V. Apostolopoulos [2], L. Kunz [3], J. Beneš [4]

[1] *Institute of Solid Mechanics, Mechatronics and Biomechanics, Faculty of Mechanical Engineering, Brno University of Technology, Technická 2896, 616 69 Brno, Czech Republic*

[2] *First Department of Orthopaedic Surgery, St. Anne's University Hospital and Faculty of Medicine, Masaryk University, Brno, Czech Republic*

[3] *Institute of Physics of Materials, Czech Academy of Sciences, Brno, Czech Republic*

[4] *BT Medical Engineering, Inc, Rozsochy, Czech Republic*

[a] *gabriel.cabaj@vutbr.cz*

Abstract: Nowadays, optical strain measurement using the digital image correlation (DIC) method is a frequently used experimental analysis method in various engineering fields. The main advantage is that it is a non-invasive method, therefore it is possible to analyze areas on structures that could be inaccessible to conventional contact methods. This paper aims to present various applications of the use of the digital image correlation method in biomechanical applications and specific case studies are presented using a commercial software Alpha from X-Sight company.

Keywords: biomechanics; strain analysis; digital image correlation; optical measurement; experimental modelling.

1 Introduction

Several experimental analyses using the DIC in the field of biomechanics are presented in this paper. It presents the possibilities of optical strain analysis, which compared to traditional contact methods, offers wider possibilities for experimental analysis of components. Examples are given where it is necessary to analyze the displacements of individual points on a component but also to analyze the whole region and evaluate the field of the strains.

In this paper, optical deformation analyses using a commercial DIC system and X-Sight's Alpha software are presented. Depending on the type of problem to be solved, the capabilities of this software are used to evaluate the displacements or strains on the analyzed object. The purpose is to show the capability of the optical method, which comes where traditional contact methods of measuring deformations are not sufficient.

The experiments presented involve the determination of the strain distribution on an orthopaedic orthosis during a cyclic test. Later, the analysis of the composite pelvic bone model before and after the application of the fixator is described. The last two sections are dedicated to the analysis of the fixed fracture of the humerus and the compression test of implanted total knee endoprostheses in the composite tibial bone.

2 Method

The DIC method is a popular tool used today for non-contact deformation measurement. Its basic principle involves tracking the displacement of a surface by analyzing successive digital images captured during the loading of a specimen. The reference digital image of the undeformed specimen is compared with the following image of the deformed specimen so that the strains can be evaluated based on the computed displacements. Each tracked point in the image is represented by a unique set of pixels from its surroundings (subset). Due to this uniqueness, it is then possible to use digital correlation to determine the new position of this subset of pixels in the next frame and thus quantify the displacement of the point represented by this subset of pixels [1].

This method can be used either to measure displacements in the plane using a single camera or to measure displacements in 3D space using at least two cameras. If two cameras are used, the measurement in 3D space is performed by capturing the image from different views at stereo camera angles. Using triangulation, it is then possible to calculate the position of a point in space and subsequently its displacement [1].

3 Strain Analysis of 3D Printed Orthopedic Orthoses

3.1 Problem Description

Supramalleolar orthosis is a type of orthotic device that is primarily used to stabilize the ankle and foot of patients. These orthoses are designed to provide control over the ankle joint but still allow for natural movement and sufficient flexibility. An example of such a device is shown in Fig. 1. It is 3D printing technology that is the appropriate way to produce such devices. Thanks to its high flexibility, this method can be used to create personalized orthoses directly tailored to patients. In addition to the fact that 3D printing enables rapid prototyping, 3D printed materials also can exhibit much better mechanical properties than conventional materials. This significantly increases the durability and reliability of these orthoses [2–4].

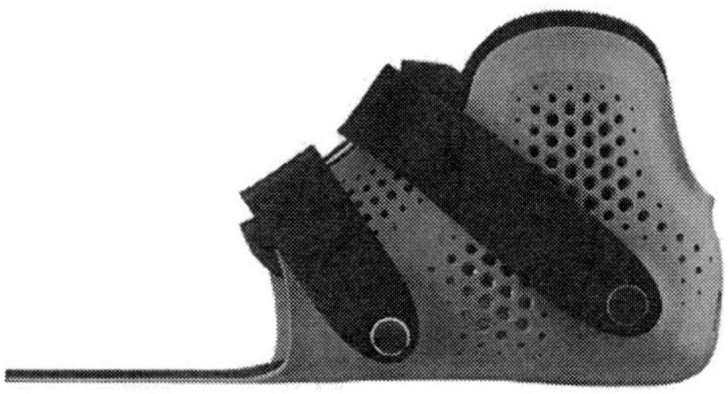

Fig. 1: Analyzed object – supra malleolar orthosis for ankle support primarily designed for young patients with pronation and low tone [5].

The main objective of this work was the experimental comparative analysis of the mechanical response of the ankle orthosis to the loads to which it is subjected during the period of use by the patient. The ankle orthosis is primarily cyclically loaded daily, while the patient walks. However, significant loading occurs mainly when the orthosis is placed on the patient's foot. In this case, there must be a significant deformation of the orthosis for it to be placed on the foot [6]. The analysis of the strain field under quasi-static loading is primarily used to identify potential critical areas and can be used for component optimization. In this work, in addition to quasi-static loading, the response of the component to cyclic loading is also investigated.

3.2 Experiment Description

In this work, a 3D measurement with two cameras was used. The cameras had 9 MPx resolution and 67 Hz frame rate. The resolution of the given camera configuration was 0.3 μm. Before the actual measurement, it was necessary to calibrate the 3D DIC system. The component was placed at 268 mm in from of both cameras, with the lateral direction perpendicular to the image.

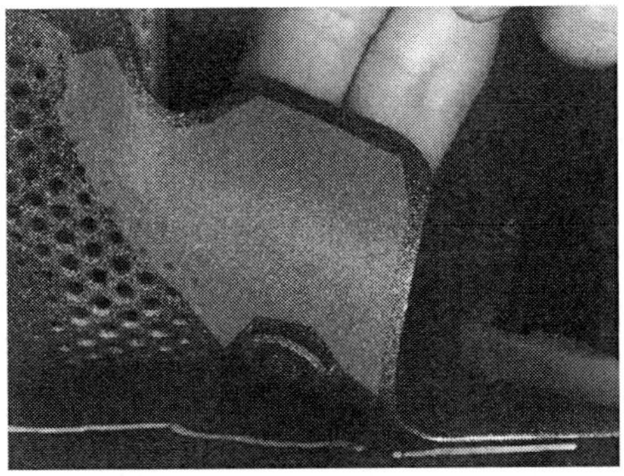

Fig. 2: Analyzed orthosis with shown area of interest.

Since the orthosis is mostly stressed when placed on the patient's foot, several orthoses with different designs were subjected to this stress, and their response was evaluated. The loading method was chosen with consideration of the orthosis loading in reality. Therefore, the orthosis was loaded as it would be when placed on the foot. Specifically, by pushing off the side of the orthosis with the patient's hand. Component loading was controlled by finger displacements in the lateral direction. As the value of the displacements was evaluated by the DIC system during the test, it was ensured that the same displacement in the lateral direction was provided for each version of the component. The loading was quasi-static.

Each component was subjected to two load cases. In the first case, it was only one load cycle with a maximum lateral displacement of 8 mm. In the second load case, the maximum lateral displacement on the component was 23 mm and 50 load cycles were performed.

3.3 Obtained Results

Overall, several different sizes of ankle orthoses were analyzed. For simplicity, only two (the biggest and the smallest) of them are presented in the comparative analysis. Since a comparative analysis was performed, the results of the field strains (first principal strain) are always presented with a consistent legend in each figure.

Comparing Figs 3 and 4, it can be concluded that the character of the strain distribution is very similar. The extreme values are localized around the rounded part – the notch. As this is a commercial product, the specific value of the results is not disclosed. The results of the analysis were used by the orthosis manufacturer to further optimize the products.

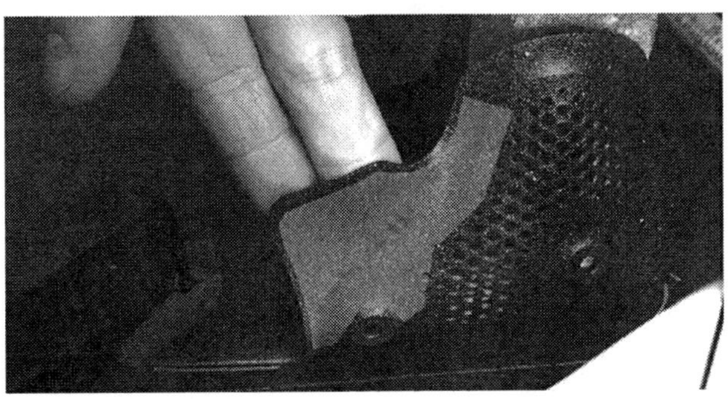

Fig. 3: First principal strain distribution on bigger ankle orthosis during static loading.

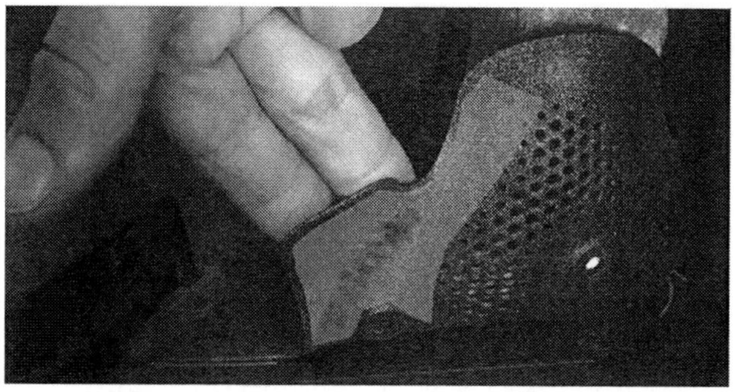

Fig. 4: First principal strain distribution on a smaller version of ankle orthosis during static loading.

Another interesting was the response of the orthoses to cyclic loading. Under full load, the character of the deformation was like that of the quasi-static loading method – see Fig. 5. After unloading under cyclic loading, clear areas of plastic strain are visible. These areas may be critical concerning fatigue, the areas are visible in Fig. 6.

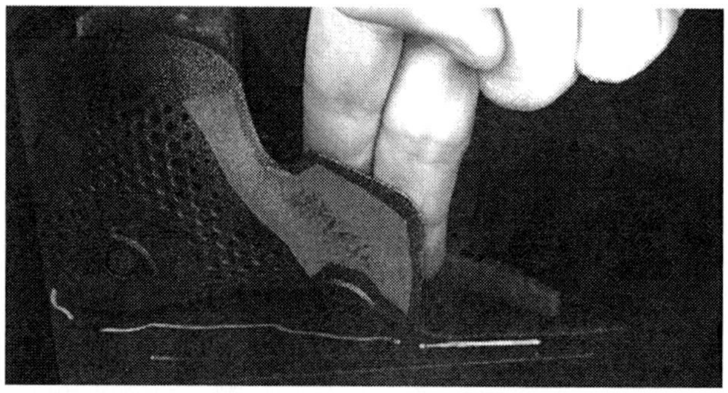

Fig. 5: First principal strain distribution on a bigger version of ankle orthosis during cyclic loading.

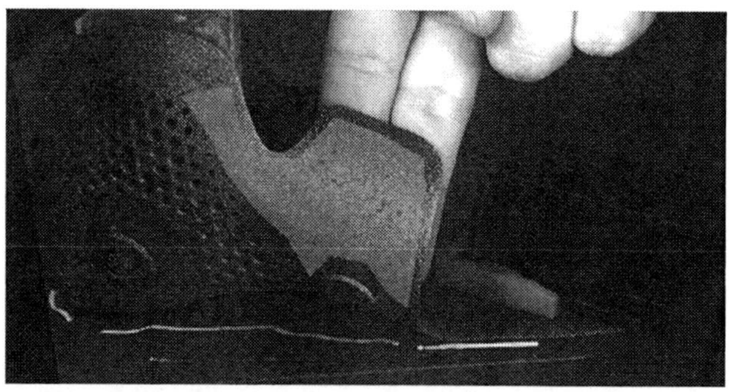

Fig. 6: Residual strain region on bigger version of ankle orthosis after cyclic loading.

4 Analysis of Strain and Displacement on a Composite Model of the Pelvic Bone

4.1 Problem Description

When the pelvic bone is injured and damaged, especially in the area of the symphysis, in common practice, it is then surgically fixed by the addition of a fixator.

The problem that was analyzed in this case can be divided into two parts. In the first case, the effect of the addition of the fixator on the distribution of the stress in the pelvis compared to the undamaged pelvis was analyzed. This is particularly important from the point of view of the patient's recovery, to answer the question of whether a fixed pelvis would be more susceptible to damage outside the region of the fixator.

In the second case, an analysis of the stress on the fixator after its application is required. This is because of its optimal design with regard to the risk of damage, especially by fatigue fracture. What is also analyzed is the effect of fixation – the number of screws [7].

4.2 Experiment Description

The experiment was carried out on an electrodynamic INSTRON E10000 testing machine. The axis of the loading force was the hip joint and the junction of the pelvis with the sacrum – see the picture below.

The whole experiment consisted of several load conditions. The response to quasi-static loading of the pelvis, with different magnitudes of loading force, was analyzed. The magnitude of the force was chosen with the consideration that the patient in recovery may start loading the pelvis gradually after surgery. Thus, each of these loading levels was evaluated.

In the next phase, the force with a different amplitude from the quasi-static test was again applied, but the loading in this case was done cyclically with a frequency of 0.5 Hz and 1 Hz.

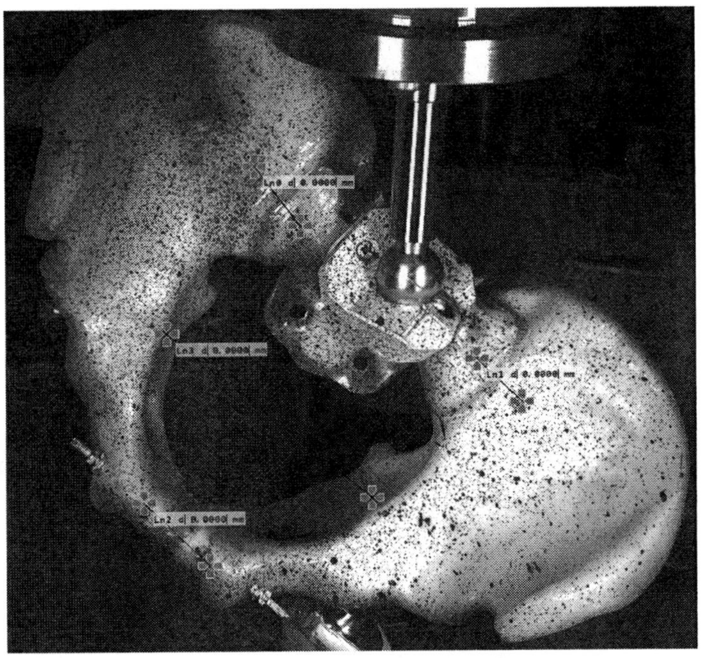

Fig. 7: Pelvic bone loading in a universal testing machine and identified measured positions.

4.3 Obtained Results

Based on the measured values of the change in individual lengths between the points shown in Fig. 7, the global behavior of the undamaged and fixed pelvic bone was analyzed.

In the next part of the experiment, the stress on the surface of the fixator (Fig. 8) was evaluated under static and cyclic loading. This was done by measuring the strains on the fixator using the DIC and converting these strains to the corresponding stresses based on the material's value of Young's modulus of elasticity. The obtained data are essential for the prediction of the fatigue life of the fixator for its design modifications.

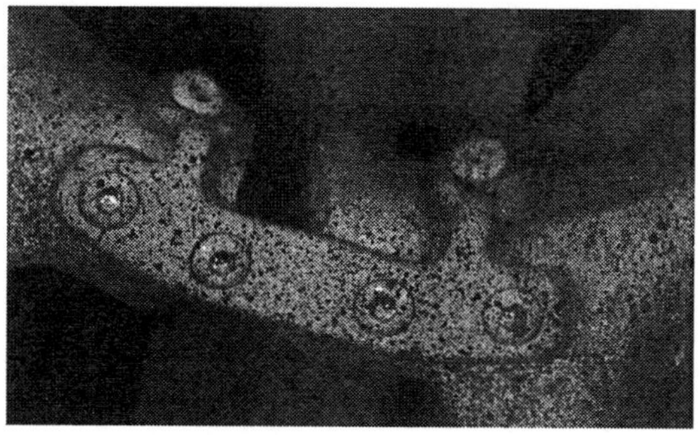

Fig. 8: Analyzed fixator – detailed view.

5 Analysis of Displacements after Fixation of Humerus Fracture

5.1 Problem Description

In a fracture of the humerus between the bone diaphysis and epiphysis, it is subsequently fixed with a fixator that is inserted into the axis of the bone. In this case, the aim was to determine how the humerus behaves with a fixed fracture. Specifically, whether there is significant relative motion between the humerus head and the diaphysis of the humerus. Subsequently, the influence of the type of fixation was also analyzed.

5.2 Experiment Description

The humerus was loaded with a certain force in a universal testing machine with the help of a special fixture that simulated the 90° abduction that occurs when raising the arm. Thus, 500 cycles were performed and the motion of the points shown in Fig. 9 was analyzed. In this case, DIC was only used to analyze the displacements of individual points from which the relative motion of the two parts, the humerus head and the diaphysis, could be determined.

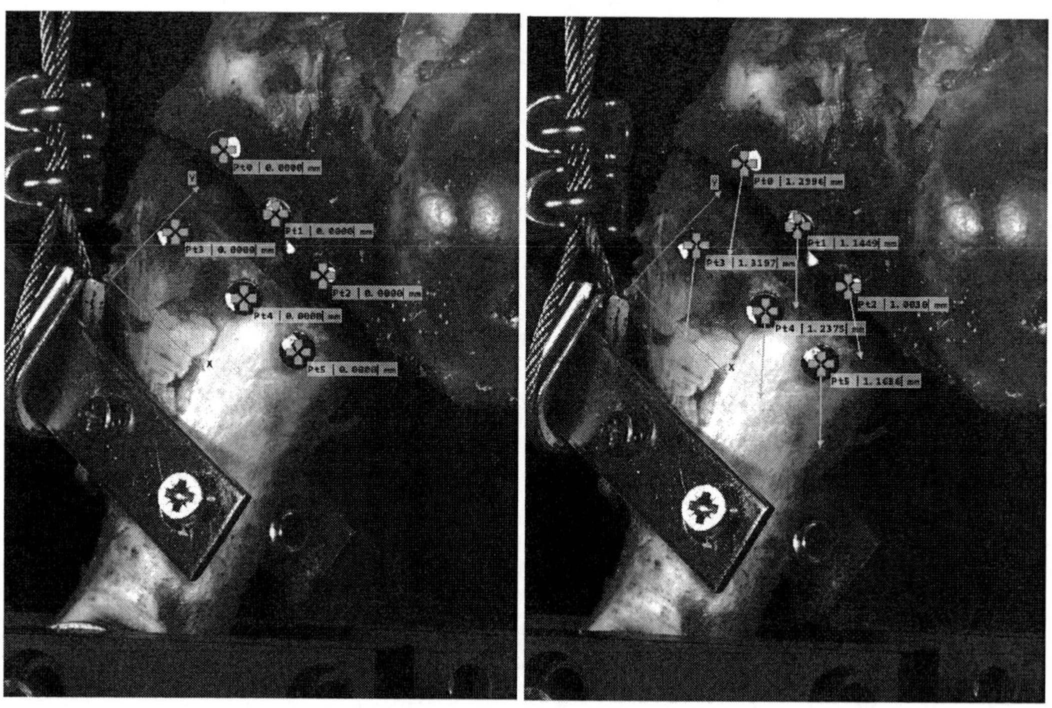

Fig. 9: Analysis of deformations in fixed humerus fracture. Relaxed state on the left and loaded state on the right. Displacements on the capitellum and the bone are shown.

5.3 Obtained Results

From the subsequent processing of the displacements of the individual points, it was possible to assess the relative movement of the two mentioned parts qualitatively and quantitatively. These results are important for the clinician in the choice of the implant and the method of its placement. Due to confidentiality, specific results are not reported.

6 Analysis of the Tibial Bone in the Compression Test

6.1 Problem Description

The final treatment of gonarthrosis involves the implantation of a total knee endoprosthesis that interacts with the tibial bone. The experimental analysis using DIC aimed to assess the effect of different types of implants on the mechanical response of the cortical bone tissue model to the composite tibial bone. This experiment was also part of a wider research with computational modelling using the finite element method [8–10].

6.2 Experiment Description

The tibia with implanted tibial part of total knee replacement was loaded using a femoral component in compression. Two different total knee prostheses were analyzed in this way. The setup of loading is presented in Fig. 10 [8–10].

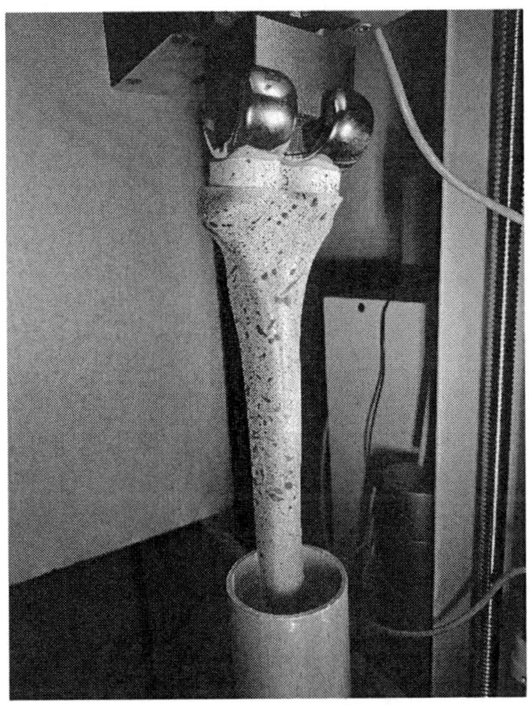

Fig. 10: Compression test of the tibial bone with the applied total knee replacement [8].

6.3 Obtained Results

Using Alpha software, the field of strain on the surface of the cortical bone tissue in the proximal tibial bone area was evaluated. Together with computational modelling, these results can be used to assess the impact of different types of implants [8–10].

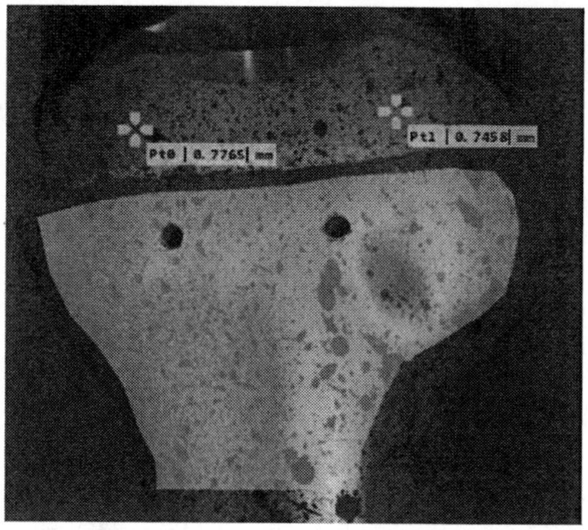

Fig. 9: Strain distribution on loaded tibial bone [8].

7 Conclusion

In this paper, several experimental analyses in the field of biomechanics using the digital image correlation method have been described.

Using the DIC system and Alpha software were used to analyze the mechanical response of the orthotic orthoses and provide the manufacturer with information that can be used in the further development of these orthoses.

In the optical analysis of displacements on the pelvic bone, the results can be used to assess the effect of the addition of a fixator on changing the character of the strain on the pelvic bone. Further, this analaysis enables us to determine the stress state on the fixator. This knowledge is crucial for the prediction of the fixator lifetime.

Analysis of the influence of the fixator was also performed in the case of humerus fracture. Here, the relative movement between the humerus head and the diaphysis after implant placement was analyzed.

The last case study presents an analysis of the strain distribution on the surface of the cortical tissue of the tibial bone during the application of different types of total knee replacements.

This paper aims to present the wide possibilities of using the DIC system. The analyses described here would be difficult to perform using conventional contact methods.

References

[1] J. Górszczyk, K. Malicki, T. Zych, Application of Digital Image Correlation (DIC) Method for Road Material Testing, Materials 12 (2019) 2349, https://doi.org/10.3390/ma12152349.

[2] D. Al-Bayati, A.M. Kadhim, The Application of Supra_Malleolar Orthosis (SMO) in Iraq: Design and Fabrication Approach, IOP Conference Series: Materials Science and Engineering 1094 (2021) 012045, https://doi.org/10.1088/1757-899X/1094/1/012045.

[3] D. Popescu, F. Baciu, D. Vlăsceanu, R. Marinescu, D. Lăptoiu, Investigations on the Fatigue Behavior of 3D-Printed and Thermoformed Polylactic Acid Wrist–Hand Orthoses, Polymers 15 (2023) 2737, https://doi.org/10.3390/polym15122737.

[4] R. Kumar, S.K. Sarangi, 3D-Printed Orthosis: A Review on Design Process and Material Selection for Fused Deposition Modeling Process, in proc: Advances in Materials Processing and Manufacturing Applications (2021), A. Patnaik, E. Kozeschnik, V. Kukshal (Eds.), Springer Singapore, Singapore, 531–538, https://doi.org/10.1007/978-981-16-0909-1_55.

[5] NEWGENERATION3D PrintedSMO, Piro (n.d.), available from: https://www.piroafo.com/smo/.

[6] M. Alvarez-Camacho, V. Alvarez-Guevara, C. Galvan Duque Gastelum, D. Flores Vazquez, G. Rodriguez-Reyes, F.M. Sanchez-Arevalo, Use of three-dimensional digital image correlation to evaluate mechanical response of prosthetic systems, Prosthetics & Orthotics International 41 (2017) 101–105, https://doi.org/10.1177/0309364616637956.

[7] J. Beneš, L. Kunz, J. Liška, J. Šrám, M. Krbec, V. Džupa, M. Křivohlávek, T. Návrat, Determination of stress and displacement on human composite pelvis under static and dynamic loading, Proceedings of the Institution of Mechanical Engineers, Part H 236 (2022) 951–959, https://doi.org/10.1177/09544119221099805.

[8] V. Apostolopoulos, Biomechanické a klinické srovnání totální náhrady kolenního kloubu typu All-poly a Metal-backed Online, Disertační práce, Brno, Masarykova univerzita, Lékařská fakulta, 2024, Dostupné z: https://is.muni.cz/th/le1t7/. [cit. 2024-04-18], n.d.

[9] V. Apostolopoulos, T. Tomáš, P. Boháč, P. Marcián, M. Mahdal, T. Valoušek, P. Janíček, L. Nachtnebl, Biomechanical analysis of all-polyethylene total knee arthroplasty on periprosthetic tibia using the finite element method, Computer Methods and Programs in Biomedicine 220 (2022) 106834, https://doi.org/10.1016/j.cmpb.2022.106834.

[10] V. Apostolopoulos, P. Boháč, P. Marcián, L. Nachtnebl, M. Mahdal, L. Pazourek, T. Tomáš, Biomechanical comparison of all-polyethylene total knee replacement and its metal-backed equivalent on periprosthetic tibia using the finite element method, Journal of Orthopaedic Surgery and Research 19 (2024) 153, https://doi.org/10.1186/s13018-024-04631-0.

Flexural Behaviour of a Fibre-Reinforced Polymer Composite: Experiment and Simulation

J. Ezenwankwo [1,a], I. Petrikova [1], J. Zak [1]

[1]*Technical University of Liberec, Liberec, Czech Republic*
ajohnson.ezenwankwo@tul.cz

Abstract: This paper involves the investigation of the flexural behaviour of a laminate composite. The three-point bending method is used with reference to ASTM D7264/7264M – 07. The laminate is composed of a single ply with 20 layers of grass fibre as reinforcement. A stacking sequence of [0_{20}] was used. Seven samples cut out from the entire board at 20 x 200 mm were used. The Thickness of the ply was 4.45 mm with a support span of 71mm according to standard. The samples were tested until visible delamination and debonding typified by visible shear and dip in applied compression up to a value of 5.8 mm. The simulation was done using Ansys ACP-Pre (plies) tied with the Mechanical Model (supports) and the Static structural module (unification of both). The results showed good agreement with the experimental result.

Keywords: laminate; bending; compression; simulation; experiment.

1 Introduction

Laminated composites have become a material of choice in modern-day applications of engineering and technology. This is attributed to its high strength-to-weight ratio, low machining cost and possibilities for extended studies in research and development. By studying the mechanical and thermal properties of composites, engineers, technologists and scientists are suitably equipped with the right experimental, numerical and simulation data to predict, with reasonable accuracy, the behaviour of materials in real-life applications. Be it advanced manufacturing techniques, natural-synthetic fibre-reinforced polymer composites, biocomposites, the application and scientific use of composites have grown in leaps and bounds, especially in the use of advanced techniques in investigating failures in composite materials [1, 2]. For years now, applications have forayed into automotive, aerospace, wind energy industries, and new frontiers have unfolded in smart composites and nanocomposites. Although fibre-reinforced composites have their tremendous benefits, they also have their inherent drawbacks and challenges arising from the fibre itself and the finished composite at various stages of their life cycle. Problems can arise in different stages including fabrication, joining, shaping, and disposal. Other concerns arise from functionality and usage, include cost, (arising from the high energy cost of production machining), quality assurance and control, durability, design complications and complexities, moisture conditions, fire resistance, compatibility with other materials, and adherence to applicable international standards for best practices [3, 4, 5]. The use of FE methods has created new ideas in modelling the real-life behaviour and failure modes of composites – be it damage initiation, debonding and delamination of composite materials under bending load [6]. The critical natures of composite materials in use make it of utmost importance that they are prequalified and tested by standards to ensure they are, not only, fit for purpose but also have the highest level of safety and structural integrity. Nowadays there are several high-quality articles that highlights the effect of stacking sequence on the behaviour of composite materials in failure [7]. Improvements in the design of carbon fibre reinforced polymer (CFRP) have led to the suitability of applications at elevated temperatures [8] and also highly dynamic loads in terms of multiaxial and fatigue stresses.

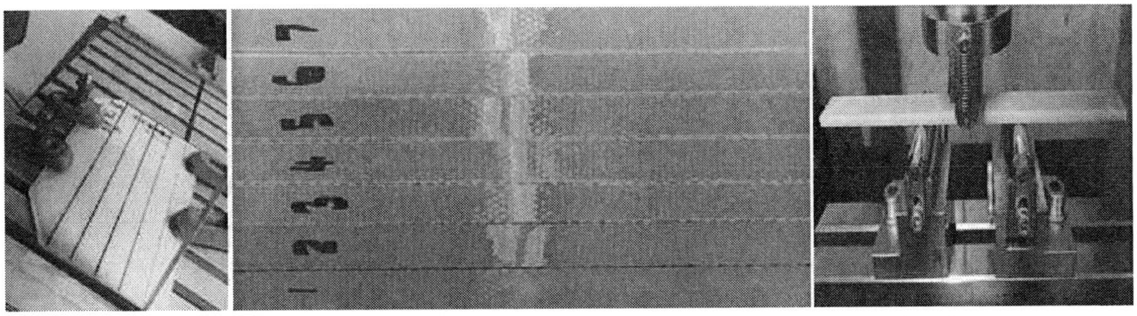

a) Laminate cutting b) Sampling and measurement c) Bending test setup

Fig. 1: Composite laminates cutting, sampling and setting up on the *Tira 2810* test bed.

2 Sample Geometrics and Experimental Setup

The fibres are cut out 250 x 250 mm. The matrix is epoxy resin with hardner mixed at a ratio of 2:1 - exactly 100:50 ml or cm^3. The classic nylon-bleeder-separator- nylon arrangement is used to set up the components before manual roller action. The arrangement is then loaded inside the curing machine for 60 minutes at a compression force and temperature of 20 kN and 75 °C respectively before cooling to ambient temperature inside the chamber. The samples are then cut according to standards as shown in Tab. 1. Seven samples were tested for each composite type.

Tab. 1: Sample Geometrics and ASTM standard specimen.

Sample Type	Fibre Configuration	Laminate Thickness	Thickness-to-Span Ratio	Width	Number of Ply	Support Span
S1	$[0_{20}]$	4.45 mm	16:10	20 mm	20	71.00 mm
S2	$[(0_{2gl}, 0_{ry})_2]s$	3.15 mm	16:10	20 mm	12	71.00 mm
S3	$[0_{16}]$	3.40 mm	16:10	20 mm	16	71.00 mm

The experimental setup is setup according to ASTM D7264/7264M – 07. A trial sample is loaded to the Tira 2810 test bed to decide at what lateral deflection visible damage in the form of surface cracks and debonding. This value is recorded. To ensure uniformity, the machine is then preset to apply this downward bending force as prescribed displacement in all tests. The value in 5.80 mm in the downward y-axis. Data for displacement and force were recorded.

3 Building the Simulation in Ansys ACP-Pre

The process involves creating two parts of the entire structure, then linking the system of ACP-pre (laminate thickness and fibre orientation), mechanical model (modelling the supports and plunger) and static structural (assembly of ACP-pre and support structure) as shown in Fig. 2.

The material data from experiment are used to determine the Young's modulus and Poisson ratio for the elastic region. The plastic region is fed in as raw data into Ansys as a multilinear kinetic hardening model to deduce the relevant constants.

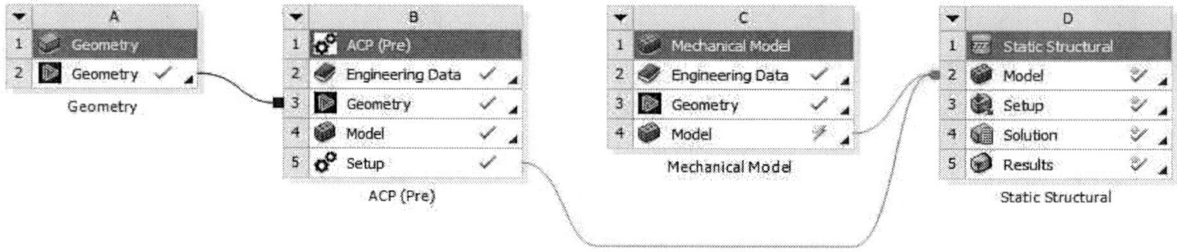

Fig. 2: Simulation layout.

The required boundary conditions and contact behaviour at touching region are created. Contact is set to simple friction with a coefficient of 0.1 between the composite and the indentor whereas it is ideally set to bonded between the supports and the lower part of the composite. The prescribed displacement is set as 5.8 mm in the downward y-direction as in the experiment. Figure 3 shows the basic schematic of the structural layout of the system. The diameter of the hemispherical supports and plunger were measured at 15 mm in the lab and implemented in modelling CAD model of the setup.

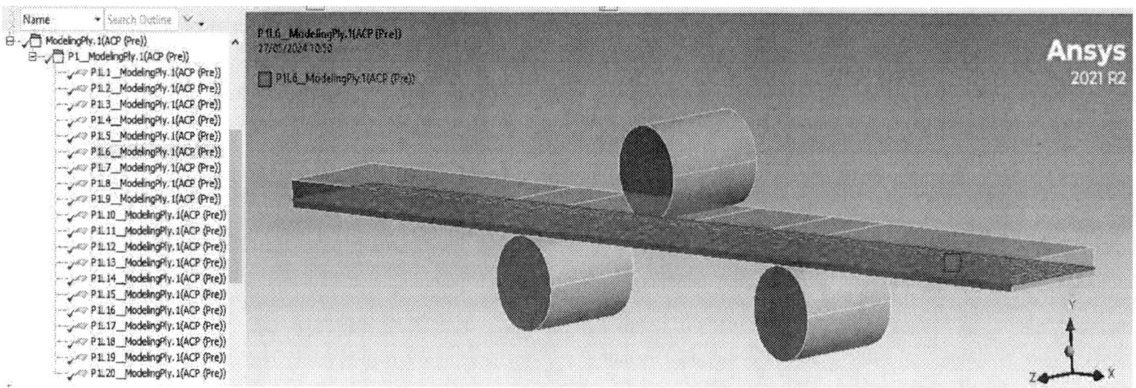

Fig. 3: Schematic of layers in ACP-pre.

4 Results and Discussion

The force-displacement data from each group of S_1, S_2 and S_3 are first averaged according to standard. Both engineering and true stress and strain were deduced for S_1, S_2 and S_3. The Young's modulus deducted from the computation of experimental data is shown in Tab. 2. The Young's modulus is calculated from the Eqs. (1) and 2 show the stress and strain respectively of a laminated composite prismatic beam in bending

$$\sigma = \frac{3PL}{2bh^2} \tag{1}$$

$$\varepsilon = \frac{6\delta h}{L^2} \tag{2}$$

where $\delta =$ mid-span deflection (mm), L = support span (mm) (determined by the suing the ASTM 16:1 span:thickness ratio), h = thickness of beam (mm).

Specimen	Young's modulus (GPa)
S_1	21.195
S_2	18.737

S_3	25.356

Figure 4 shows the plots of stress-strain and force-displacement. The modulus of elasticity is derived from the linear region of the curve. Data from the plastic region is fed into an in-built multilinear kinetic model.

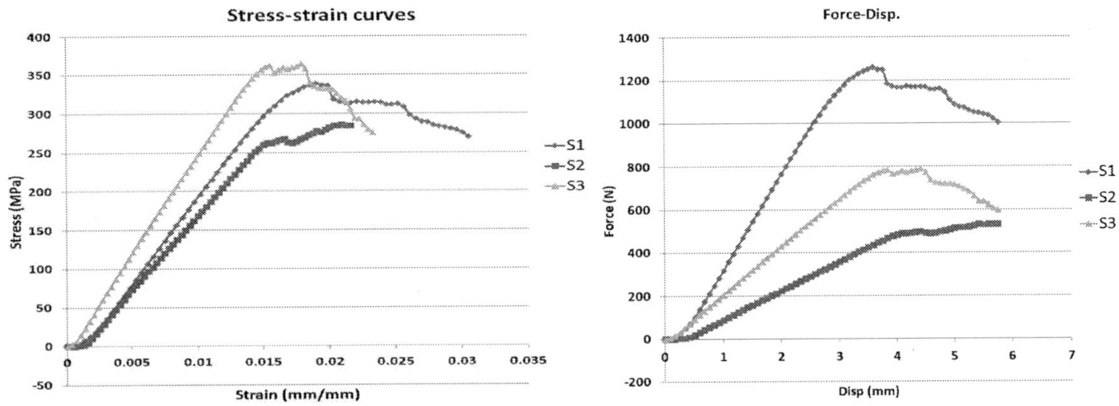

a) Stress-strain plot for the samples b) Force-displacement

b) Fig. 4: Plots of the result from the flexural tests for all samples.

The test console produces the force, displacement and time data which are used to compute the stress and train after averaging for samples of each group. Figure 5 shows the distribution of Young's modulus for the three samples.

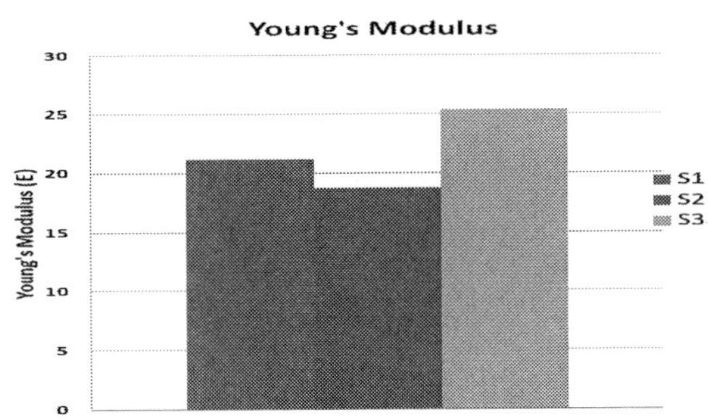

Fig. 5: Plots of Young's modulus (E_l) for the different samples.

A comparison is made of the data collected from Ansys post-processing. The plots are shown in Fig. 6 for the S_l sample. The differences in the experimental and simulation could be attributed to the difficulty in manufacturing the composites to an ideal physical state. Geometrical nonlinearities are also another factor.

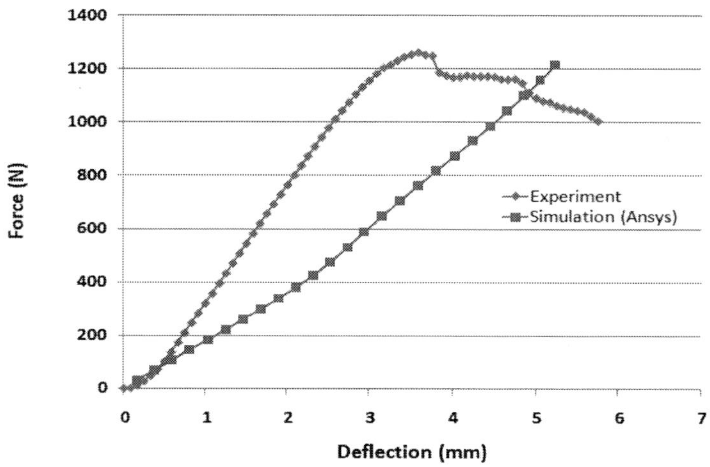

Fig. 6: Force-displacement plot for (S1).

5 Conclusion

The finite element option provides a more convenient, time-saving, yet complex, method of building composite plies Ansys ACP-pre. It also provides a better approach in carrying out both flexural and tensile tests while helping to eliminate such physical ordeals as slippage at the clamps and supports during repeated experimental testing. It also gives the opportunity to appreciate that ply orientation plays a key role in the behaviour of laminated composites. The zero-degree ply has been shown to have the greadest breaking axial force. This is particularly useful when designing components fit-for-purpose. Further work would be in high-speed impact test of the laminated composite with emphasis on the effect of material type (glass fibre, rayon, glass mat).

Acknowledgement

This publication was written at the Technical University of Liberec as part of the project "Research of advanced materials, and application of machine learning in the area of control and modeling of mechanical systems" nr. SGS-2022-5072 with the support of the Specific University Research Grant, as provided by the Ministry of Education, Youth and Sports of the Czech Republic in the year 2024.

References

[1] Q. Shen, M. Omar, S. Dongri, Ultrasonic NDE Techniques for Impact Damage Inspection on CFRP Laminates, Journal of Materials Science Research 1(1) (2012) 34–36, DOI: 10.5539/jmsr.v1n1p2.

[2] S. Thabet, Y. Jasim, T. Thabit, Critically Evaluate the Capabilities of Ultrasonic Techniques Used for Tracing Defects in Laminated Composite Materials, International Journal of Engineering and Applied Sciences 10(3) (2018) 237–251, http://dx.doi.org/10.24107/ijeas.465749.

[3] R. N. Shama, T. G. A. Simha, K. P. Rao, G. V. V. Ravi Kumar, Carbon Composites are Becoming Competitive and Cost Effective, Infosys: NYSE, 2018.

[4] M. G. Bader, Selection of Composite materials and Manufacturing Routed for Cost-Effective Performance, Composites Part A: Applied Science and Manufacturing 33(7) (2002) 913–934, https://doi.org/10.1016/S1359-835X(02)00044-1.

[5] C. Lester, S. Nutt, Composite Materials: Advantages and Cost Factors, PhD, Material Science, 2018.

[6] H. Ullah, A. R. Harland, T. Lucas, D. Price, V. V. Silberschmidt, Computational Materials Science 52 (2012) 147–156.

[7] A. Azzam, W. Li., An Experimental Investigation on the Three-point Bending Behavior of Composite Laminate, IOP Conference Series: Materials Science and Engineering 62 (2014) 012016.

[8] R. M. Abdalrahman, Flexural Behaviour of CFRP Laminate at Elevated Temperature, Kurdistan Journal of Applied Research 2(3) (2017) 330–334, https://doi.org/10.24017/science.2017.3.26.

Impact of Direct Ageing on Mechanical Properties and Residual Stresses of Maraging 300 Steel Manufactured by SLM

R. Halama [1,a], M. Kořínek [1], J. Čapek [2], K. Trojan [2], J. Hajnyš [1]

[1] *VŠB-Technical University of Ostrava, 17. listopadu 2172/15, 708 00 Ostrava-Poruba, Czech Republic*
[2] *Czech Technical University in Prague, Trojanova 13, 120 00 Prague, Czech Republic*
[a] *radim.halama@vsb.cz*

Abstract: This paper describes the main results of research conducted on samples manufactured by selective laser melting of maraging steel 300 powder. The results of the tensile tests helped us to construct maps of yield and ultimate strength as a function of temperature and time hold of direct ageing of the material. Furthermore, X-ray diffraction was used to determine the decrease of residual stresses with time during heat treatment under the lowest temperature. The magnitude of residual stresses in the axial direction of the sample was reduced by 60% by heat treatment.

Keywords: additive manufacturing; residual stresses; X-ray diffraction; maraging steel; heat treatment.

1 Introduction

Selective laser melting (SLM) is a technology that falls into the group of additive manufacturing (AM). SLM is a promising technique for consolidating metal powder and presents a plethora of opportunities for the creation of parts. Using a scanning energy source, it entails selectively melting sections of a thin, flat powder bed layer by layer (20–100 μm thick) to create 3D parts [1]. Complex residual stresses (RS) distribution is produced during the SLM process when regions are heated in accordance with the scan method, allowing parts to heat and cool independently. The application of heat treatment (HT) prior to removing the part from the build plate and removing supports is an efficient approach to reduce RS in printed parts. To prevent microstructural alterations, the temperature for HTs must be carefully chosen; otherwise, the high strength of the material may be lost.

The maraging steel 300 (1.2709 or MS300) has high strength, excellent toughness, and weldability, is based on precipitation hardening and ranks among a nickel-iron alloy. The reason for the increased strength of maraging steel is that the intermetallic nanoparticles Ni_3(Mo, Ti) and Fe_2Mo precipitate during ageing [2]. Steel is used extensively in the automobile, aerospace and tooling industries [3]. To modify the microstructure and improve the mechanical properties, conventional HT is commonly applied after printing by the SLM method. The HT usually consists of solution treatment, solution ageing, and direct ageing [4, 5]. In recent years, several studies have been conducted on maraging steels and their HT. The results show that in the as-built condition, the tensile strength values are around 1200 MPa and the ductility is 12% [6]. Applying direct ageing at 480°C and ageing for 5 h, the ultimate tensile strength values of 2217 MPa can be achieved [7]. When combining both HTs, namely solution treatment at 820°C for 1 h and ageing at 460°C for 5 h, the ultimate tensile strength of 2033 MPa was obtained, as the tensile ductility decreased from 10% to 5% [8].

The impact of direct ageing at various temperatures and different holding times on the mechanical properties of MS300 samples prepared with SLM has been reported in a preliminary study [9]. In this contribution, we precise the results of mechanical properties from new tensile tests and the impact of HT on residual stresses in vertically printed specimens made of MS300 is investigated

2 Material and Methods

2.1 Parameters of Additive Manufacturing Process

The standard chemical content of the metallic powder is available in the product data sheet [10]. All specimens for mechanical testing were printed using the Renishaw AM500E SLM printer. The print of the specimens was done in the Z direction, which is the vertical orientation of the specimen axis. Table 1 lists the primary printing parameters that are used in the AM process. To avoid oxidation, argon with a purity of 5.0 was poured into the chamber during production. To reduce residual stresses in the AM process, the building platform was heated to 170°C.

Tab. 1: Printing parameters used in the SLM process.

3D printing strategy	Laser power [W]	Laser rate [mm · s^{-1}]	Layer thickness [μm]
"meander"	400	650	40

2.2 Postprocessing

In this study, we present new results from tensile tests considering HT: 490°C, 520°C, 840°C, 900°C (always 1, 2, 6, and 10 hours). A chamber furnace LH 120/12 from Nabertherm GmbH was used for the heat treatment of the tested samples. This furnace has a power of 12 kW and reaches temperatures of up to 1200°C. Individual samples were placed in the oven for a predetermined time and temperature. The initial temperature was 22°C, from which it increased to a predetermined temperature at a rate of 600°C/h. After the given temperature was reached, there was a holding time at the heat treatment temperature. After the holding time had elapsed, the samples were withdrawn, and cooling continued slowly in the air.

2.3 Tensile Testing

All tensile tests were performed using the LabControl 100 kN / 1000 Nm hydraulic testing machine. The diameter of the round sample was 5 mm and the length of the straight part was 40 mm. The position rate applied in all tests was 10 mm/min. For each HT, the tensile test was performed three times, while for one sample, the Epsilon Tech 3560 biaxial extensometer was used to obtain information on the transverse strain. The disadvantage of the extensometer is the short gauge length (10 mm). Necking usually occurred outside of the measured area. Therefore, the ductility from such tests is not included in the overall evaluation. In all other tests, the axial strain was measured by digital image correlation (2×2.3Mpx CCD cameras, Alpha X-Sight software).

2.4 Residual Stress Determination

X-ray diffraction measurements to obtain RS were done on samples subjected to HT at 490°C. Macroscopic RS were determined from the circumference of the samples in the axial direction samples. The RS values were calculated from lattice deformations determined on the basis of experimental dependencies of $2\theta(\sin^2\psi)$ assuming a biaxial state of residual stress (θ is the diffraction angle, ψ the angle between the sample surface and the diffracting lattice planes). The diffraction angle was determined as the centre of gravity of the Cr$K\alpha_1\alpha_2$ doublet diffracted by the lattice planes $\{211\}$ of the α-Fe phase. The X-ray elastic constants $\frac{1}{2}s_2 = 5.76$ TPa^{-1}, $s_1 = -1.25$ TPa^{-1} were used for the stress calculation. The experimental error is the standard deviation according to the "$\sin^2\psi$" residual stress calculation algorithm.

3 Results

3.1 Mechanical Tests

The resulting mechanical properties are presented in the form of characteristic engineering stress-strain curves for particular HTs in Fig. 1a. For a comparison, contour maps are presented in Figs. 1b-d. There are two local extrema in the map of yield strength and ductility evaluated from tensile tests on heat treated samples, see Figs. 1b,c. The highest ultimate strength is obtained from the temperature of 490°C and 10 hours (Fig. 1d).

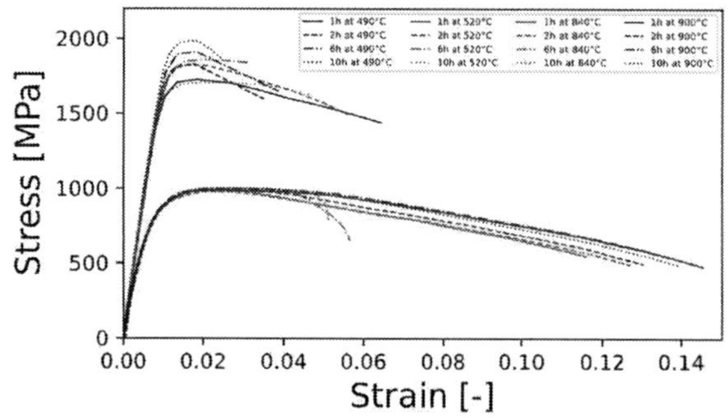

a) Engineering stress-strain curves

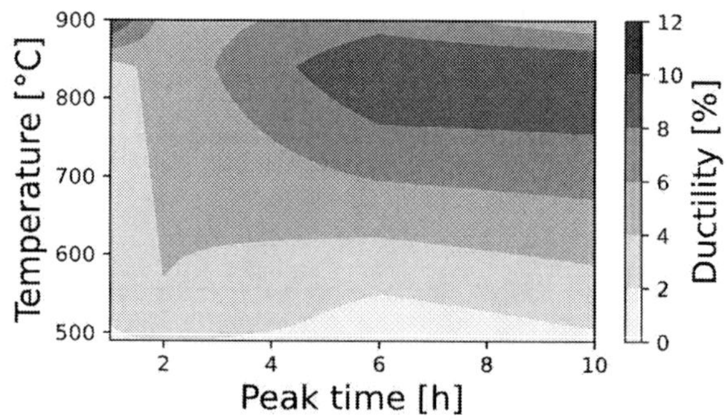

b) A map of the ductility

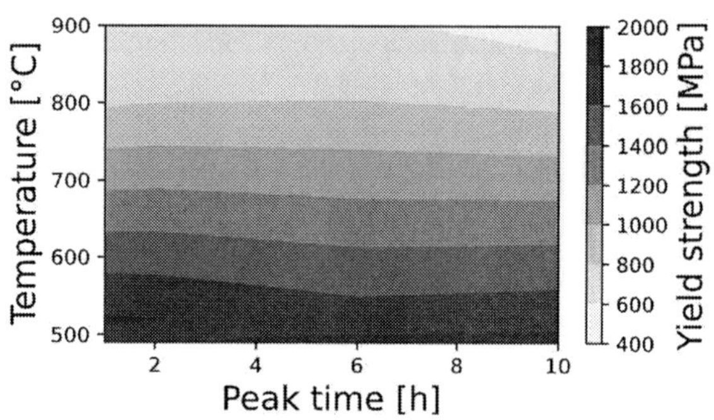

c) A map of the yield strength

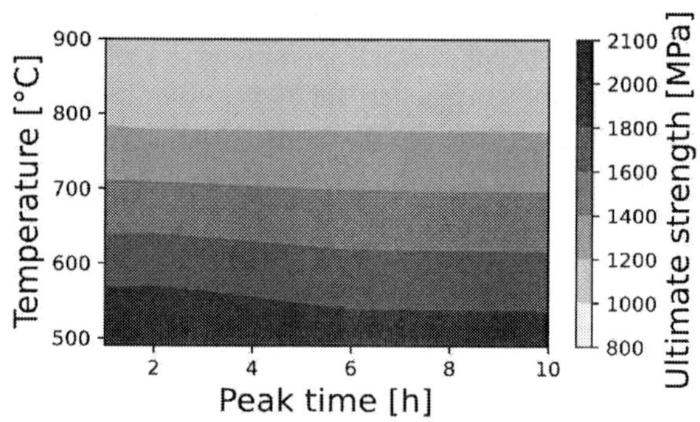

d) A map of the ultimate strength

Fig. 1: Results of mechanical tests on MS300 manufactured by SLM with specific HTs.

3.2 Residual Stresses

The temperature of 490°C was selected for the investigation of averaged RS analyzed by X-ray diffraction around the circumference of MS300 in the axial direction as a function of hold time (Fig. 2). All analyzed samples showed tensile residual stresses. As the annealing time increased while maintaining the same temperature, the average tensile residual stress of the test specimens decreased. When the time was increased to five times, there was a decrease in residual stresses by about 60%. This observation is not surprising since the degree of residual stress redistribution is a function of temperature and time. Since there was a continual decrease in adverse tensile stresses with increasing annealing time, it is not surprising that the specimen annealed at a temperature of 490°C for 10 hours exhibited the highest ultimate strength. The residual stresses add to the applied stresses and when the tensile residual stresses dominate the surface, there is effectively a reduction in ultimate strength.

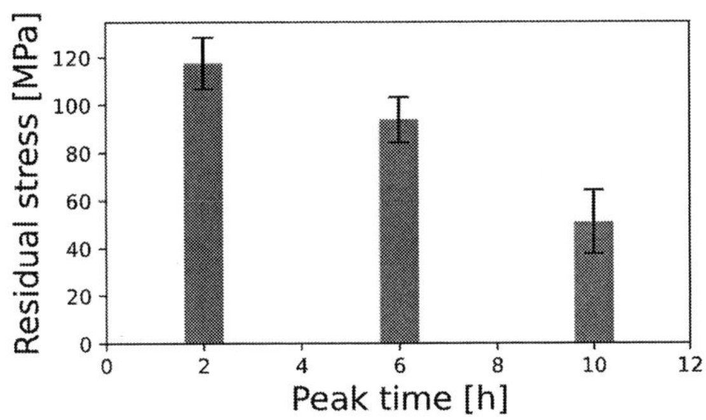

Fig. 2: Results of residual stress on MS300 SLM resulting from HTs.

4 Conclusion

In total, sixteen HT conditions, as a combination of four temperatures and four time holds, have been investigated in this study conducted on MS300. The best case from the strength increase point of view has been selected for the evaluation of the decrease in RS due to HT. The magnitude of the axial component of residual stresses decreased to 40% of its initial value. The findings of the study indicate that RS cannot be completely eliminated by 10 hours of classic direct aging. However, the specimen with the lowest tensile residual stresses exhibited the highest ultimate strength. A technique using uneven cooling developed for AlSi10Mg [11] can eventually help to reduce residual stresses in printed components.

Acknowledgement

This work is an output of the project with financial support from the Czech Science Foundation under registration no. 23-04724S. The work of CTU staff was supported by the Grant Agency of the Czech Technical University in Prague, grant No. SGS22/183/OHK4/3T/14.

References

[1] R. Halama et al., Cyclic plasticity of additively manufactured metals, in book: Cyclic Plasticity of Metals: Modeling Fundamentals and Applications, Amsterdam, Elsevier, 2022, 397–433, https://doi.org/10.1016/B978-0-12-819293-1.00022-X.

[2] T. Allam et al., Tailoring the nanostructure of laser powder bed fusion additively manufactured maraging steel, Additive Manufacturing 36 (2020) 101561, https://doi.org/10.1016/j.addma.2020.101561.

[3] E.-A. Jagle et al., Precipitation and austenite reversion behavior of a maraging steel produced by selective laser melting. Journal of Materials Research 29 (2014) 2072–2079, https://doi.org/10.1557/jmr.2014.204.

[4] Y. Bai et al., Effect of heat treatment on the microstructure and mechanical properties of maraging steel by selective laser melting, Materials Science and Engineering 760 (2019) 105–117, https://doi.org/10.1016/j.msea.2019.05.115.

[5] J. Song et al., Effect of heat treatment on microstructure and mechanical behaviours of 18Ni-300 maraging steel manufactured by selective laser melting, Optics & Laser Technology 120 (2019) 105725, https://doi.org/10.1016/j.optlastec.2019.105725.

[6] J. Suryawanshi et al., Tensile, fracture, and fatigue crack growth properties of a 3D printed maraging steel through selective laser melting, Journal of Alloys and Compounds 725 (2017) 355–364, https://doi.org/10.1016/j.jallcom.2017.07.177.

[7] K. Kempen et al., Microstructure and mechanical properties of selective laser melted 18Ni-300 steel, Physics Procedia 12 (2011) 255–263, https://doi.org/10.1016/j.phpro.2011.03.033.

[8] J. Mutua et al., Optimization of selective laser melting parameters and influence of post heat treatment on microstructure and mechanical properties of maraging steel. Materials & Design 139 (2018) 486–497, https://doi.org/10.1016/j.matdes.2017.11.042.

[9] A. Kvita, Research into the Influence of Heat Treatment on the Mechanical Properties of the MS300 Material Produced by the SLM Method, Diploma thesis, Ostrava, Faculty of Mechanical Engineering, VŠB-TU Ostrava, 2022. (in Czech)

[10] Maraging steel M300 powder for additive manufacturing, RENISHAW [online], [cit. 2024-05-21], available at: https://www.renishaw.com.

[11] C. H. Lim et al., Novel method of residual stress reduction for AlSi10Mg manufactured using selective laser melting without compromise of mechanical strength, Virtual and Physical Prototyping 18 (2022) e2131583, https://doi.org/10.1080/17452759.2022.2131583.

Methodology and Determination of Parameters for Modeling Brittle Adhesive

L. Horák[1,*], J. Krystek[1]

[1] *Department of Mechanics, Faculty of Applied Sciences, University of West Bohemia, Univerzitní 8, 301 00 Pilsen, Czech Republic*

* *horakl@kme.zcu.cz*

Abstract: In the numerical simulation of adhesive joints, a significant challenge lies in obtaining reliable experimental data due to the wide variety of adhesives available, each with distinct material properties and different material models. This study focused on testing the methodology for obtaining mechanical parameters using bulk adhesive specimens. The experiments were conducted on a brittle adhesive, specifically Araldit AV 138+HY 998, which was selected as a representative material for structural adhesives. The study also evaluated fracture parameters, including the strain energy release rate, an essential parameter for predicting the fracture behavior of adhesive joints. By focusing on a specific adhesive material, this study aimed to obtain its mechanical parameters and use them in numerical simulations.

Keywords: Adhesive; Mechanical testing; Bulk specimen; DCB; ENF.

1 Introduction

Mechanical testing of adhesives is crucial in ensuring bonded assemblies' reliability and durability. Bulk specimens are commonly used for this purpose, as they provide a convenient and standardized way to measure the mechanical properties of adhesives.

Bulk specimens are typically prepared by casting or molding the adhesive into a standardized shape, such as a dog–bone or cylinder. The specimen is then subjected to various mechanical tests, including tensile, compression, and shear tests. These tests measure the stress-strain behavior of the adhesive, which is used to calculate important mechanical properties such as the modulus of elasticity, yield strength, and ultimate tensile strength. [1]

One of the main advantages of using bulk specimens for mechanical testing is that they allow for measuring the adhesive's properties independent of the substrate material. This is important because the substrate material can influence the mechanical properties of the adhesive, and testing the adhesive alone provides a more accurate measure of its intrinsic properties.

However, it is essential to note that the mechanical properties of bulk specimens may not always correlate directly with the performance of the adhesive in real-world applications. This is because the mechanical behavior of adhesives can be influenced by various factors, including the surface preparation of the substrate, the thickness and geometry of the bond line, and the environmental conditions during application and use. [2]

To address these challenges, researchers have developed a range of advanced testing methods that can more accurately simulate the conditions experienced by adhesives in real-world applications. These methods include dynamic mechanical analysis, fracture mechanics testing, and environmental testing.

All these parameters are then applicable to associated material models in numerical simulations, which, with suitable criteria for adhesive failure, can determine safe loads and the duration of these loads. [3]

2 Experiments

The thixotropic paste adhesive Araldit AV 138, with hardening component HY 998, was selected as a representative of structural brittle adhesives. It is a two-component epoxy–based adhesive that cures at room temperature and is considered appropriate for high-strength use in aggressive or warm environments. A series of experiments were conducted to obtain the mechanical parameters specified in Table 1, where appropriate parameters and tests are associated. [5] All experiments were conducted using the Zwick/Roell Z050 universal testing machine (UTM) equipped with a 50 kN load cell. The tests were performed at an ambient temperature of 23 $\pm 1°$C, and the atmospheric moisture was within the range of $50 \pm 6\%$

Since the adhesive tested in this study is brittle and elastic, no mechanical parameters for plasticity were assumed. The experiments focused on determining the elastic, fracture initiation, and crack propagation parameters, which are critical for predicting the behavior of adhesive joints under various loads and conditions.

Tab. 1: Parameters for complex numerical simulation.

Zone name	Parameter name	Symbol	Unit	Experiment
Elastic	Young's modulus	E	[Pa]	Tensile test
	Poisson's ratio	ν	[-]	Tensile test
	Shear modulus	G	[Pa]	Arcan test
Iniciation of Fracture	Tensile failure strength	σ_{f}^T	[Pa]	Tensile test
	Shear failure strength	τ_{f}	[Pa]	Arcan test
	Tensile failure strain	ϵ_{f}^T	[-]	Tensile test
Crack propagation	Strain energy release rate for mod I	G_I	[Jm^{-2}]	DCB
	Strain energy release rate for mod II	G_{II}	[Jm^{-2}]	ENF

Bulk adhesive specimens were prepared for tensile, compression, and shear (Arcan) tests. To fabricate these specimens, a thin plate of adhesive was cast in a simple mold. During the mixing process, the two components were combined in an open container, which inevitably introduced air into the adhesive mixture. To mitigate this issue and ensure uniformity, the cast adhesive plate was subjected to a vacuum pump. The vacuum treatment was varied, with exposure times ranging from 2 to 25 minutes and negative pressures between -800 and -990 mbar. After the curing process at room temperature, selected samples underwent milling to achieve precise geometry and thickness tailored to the specific experimental requirements. The dimensions of the tensile and shear specimens are outlined in Fig. 1. The Arcan sample was configured to align with the fixtures in the Arcan test setup and facilitate fracture initiation from the tip [4]. In contrast, the tensile dog–bone test specimen was crafted to accommodate a biaxial extensometer with minimal height and dimensions.

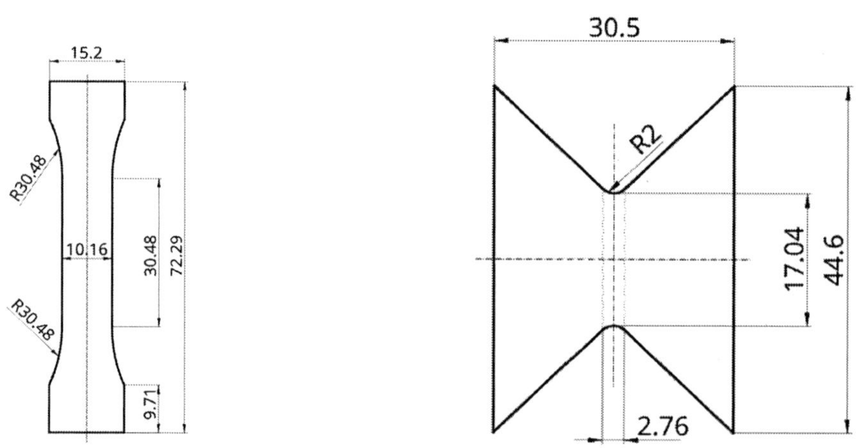

(a) Tensile sample

(b) Shear (Arcan) sample

Fig. 1: Geometry [mm] of samples used in tensile and shear experiments.

Double cantilever beam (DCB) and end-notched flexure (ENF) tests were performed to evaluate crack propagation parameters. DCB samples were made from two aluminum prismatic strips with dimensions $150 \times 20 \times 2$ mm, all dimensions are outlined in Fig. 2 (a). The prismatic strip surface was roughened with sandpaper and cleaned with an acetone-based cleaner. Initial delamination was $a_0 = 65$ mm. Load transfer was done through loading blocks, which were bonded to the specimen by the same adhesive. For one batch of samples, the specimens were placed in a vacuum chamber for 10 minutes with a running vacuum pump. This procedure aimed to assess the possibility of removing all air bubbles without altering the material properties of the adhesive.

ENF samples were made from two steel prismatic strips with dimensions $150 \times 20 \times 3$ mm, all dimensions are outlined in Fig. 2 (b). The prismatic strip surface was prepared as DCB samples. Initial delamination was $a_0 = 55$ mm. For correctly evaluating the propagation of the crack, it is crucial that any residual adhesive on the side was removed and the side was cleaned. If sandpaper is used, it is recommended that the sample be sanded in a perpendicular direction before testing. The change in material from aluminum to steel is due to the larger shear stress on the adhesive and the increased plastic deformation of the adherend.

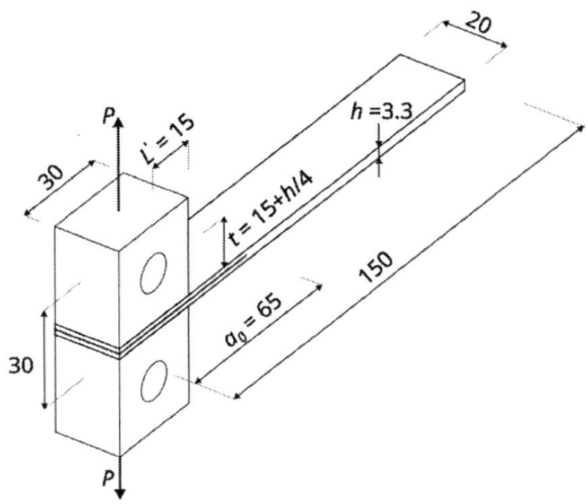

(a) DCB sample

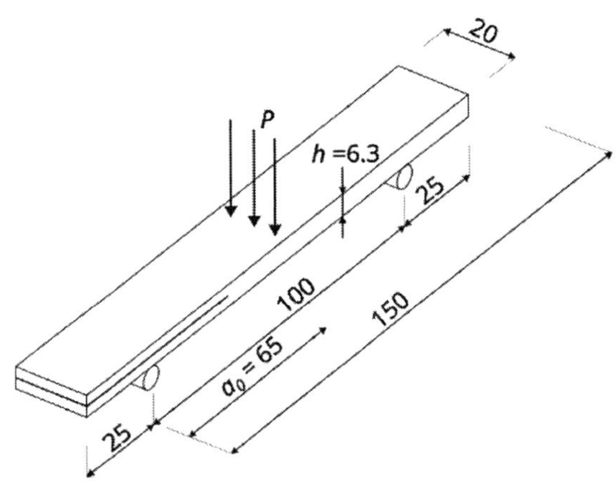

(b) ENF sample

Fig. 2: Geometry [mm] of samples used in DCB and ENF experiments.

2.1 Tensile Experiments

For evaluating Young's modulus E, Poisson's constant ν, and ultimate tensile stress σ_f^T, tensile samples were loaded with prescribed constant crosshead rate $v_T = 1$ mm/min.

In total, 11 specimens were tested, which differed in thickness h from 1 to 3 mm and slight deviation when fabricating the base plate to eliminate air in the mixture. The Young's modulus was evaluated on a range of longitudinal strain values (ϵ_L) between 0.005 and 0.015, using data from 11 test specimens. Tensile failure strength was evaluated on seven samples. The remaining samples had visible air bubbles in place of the fracture. Poisson constant was measured with a biaxial extensometer and evaluated on the same longitudinal strain ϵ_L from 0.005 to 0.015 as Young's modulus. Five samples were utilized for this evaluation. The fracture was observed in the gauge section and out of the biaxial extensometer fixture. By this, the shape and dimensions of the tensile specimen were validated. Stress–Strain curve is on Fig. 3 (b).

Due to different ways of manufacturing the adhesive plate, it was observed that some of the samples had air bubbles in place of fracture, which prematurely initiated the fracture. This was visually inspected after the test and is visible in Fig. 3 (a). These samples had approximately 35 % lower failure strength. Many samples exhibiting high failure strength did not display a completely homogeneous structure along their fracture path.

2.2 Arcan Experiments

The Arcan samples were loaded at a prescribed constant crosshead rate of $v_{Arc} = 1$ mm/min. The samples were positioned at the center using 3D–printed plates and secured to the Arcan fixture using clamps tightened

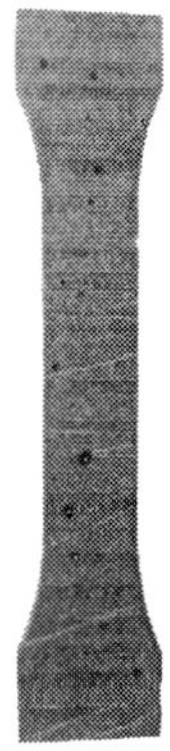

(a) Photograph of fractured sample with non–uniform structure

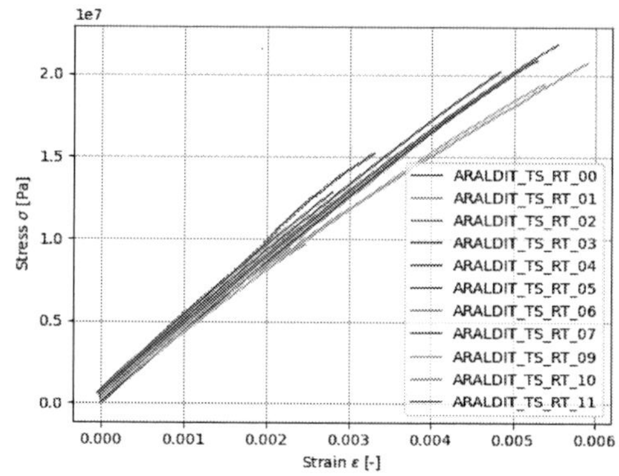

(b) Force–displacement curve for tensile experiment

Fig. 3: Tensile experiment.

by bolts. The bolts were screwed with a controlled torque of $M = 5$ N/m. The arcan experiment was performed to evaluate shear failure strength.

Force–displacement curve is on Fig. 4 (b), with visualization of fractured sample in the fixture on Fig. 4 (a). Accurate displacement was not measured, and the shear modulus was computed using Young's modulus and Poisson's constant.

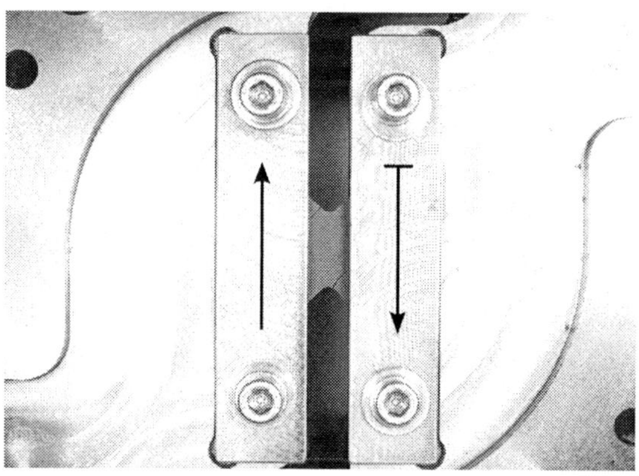

(a) Arcan sample in UTM fixture with visible cracks

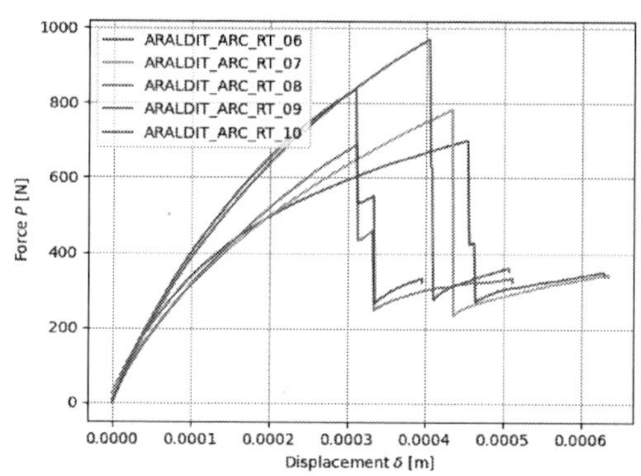

(b) Force–displacement curve

Fig. 4: Arcan experiment.

An adhesive crack initiated as expected at the location of the smallest cross-section of the sample. However, the crack propagated further to the sample. This unexpected crack propagation may have been caused by the presence of air bubbles in the adhesive. In all test specimens, two cracks were observed to form nearly simultaneously at the point of failure. The formation of these two cracks is evident in the Force–displacement curve recorded during testing.

2.3 Compression Experiments

The compression experiments were performed according to standard ASTM D695. Compression samples were loaded with prescribed constant crosshead rate $v_C = 1$ mm/min. Samples were milled from the same plate as shear and tension samples. Cylindrical samples with approximate dimensions $h = 4.8$ mm and radius $d = 12.4$ mm were used. This experiment was performed to evaluate compression failure strength. Force–displacement curve is on Fig. 5 (b).

Specimens with numbers 01 and 02 had visible air bubbles on the side of the specimen, as visible in Fig. 5 (a). The rest seems without any defect. Excluding the first specimen, the compression failure strength remained

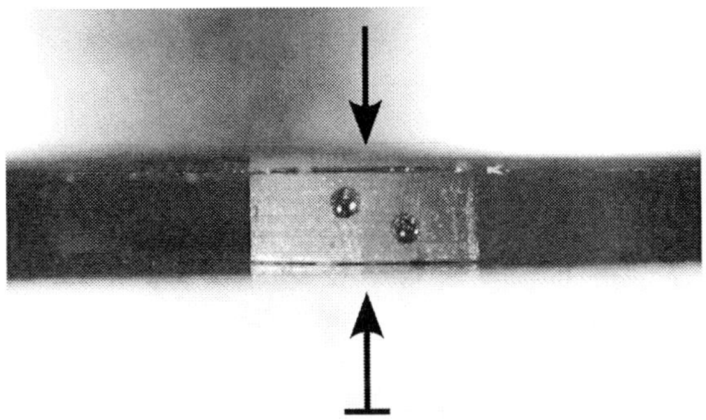

(a) Compression sample in UTM

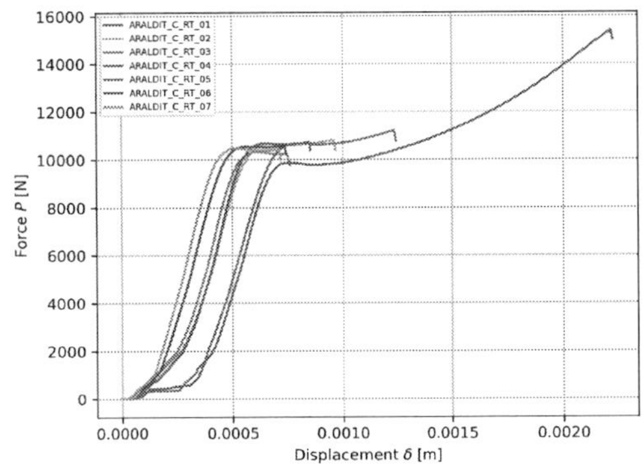

(b) Force–displacement curve

Fig. 5: Compression experiment.

consistent across all measured specimens. The first specimen, numbered 01, was measured beyond the failure, while the remaining samples were measured until the first force drop occurred.

2.4 DCB Experiments

DCB experiments were performed according to standard ASTM D5528-01. The specimen was loaded with a constant crosshead rate $v_{\text{DCB}} = 10$ mm/min. All samples were fractured cohesively, corresponding with the expected fracture mode. For further evaluation, adhesive fracture toughness for mode I was computed according to the modified beam theory as

$$G_I = \frac{3P\delta}{2B(a + |\Delta|)}, \tag{1}$$

where P is applied load during crack propagation with crack length a, δ is load point displacement, B is the width of the specimen, and Δ is a term for correction due to possible rotation, which may be determined experimentally by generating a least squares plot of the cube root of compliance as a function of delamination length.

Due to the use of loading blocks that have pins further from the top face of the specimen, correction due to stiffening is needed. Correction is computed as

$$N = 1 - \left(\frac{L'}{a}\right)^3 - \frac{9}{8}\left[1 - \left(\frac{L'}{a}\right)^2\right]\left(\frac{\delta t}{a^2}\right) - \frac{9}{35}\left(\frac{\delta}{a}\right)^2, \tag{2}$$

where L' is the distance from the loading pin to the end of the block in the direction to the opposite end of the sample, and t is the distance from the pin to the top face of the sample. To this distance, $h/4$ is added. These parameters are characterized on Fig. 2 (a). The correction is applied to compliance C as C/N when plotting compliance versus crack length. The crack length was measured from synchronized photographs taken during measurements.

In total, nine samples were tested, where samples from 1–4 were bonded without any additional steps, and samples 5–9 were let in a vacuum chamber with a vacuum pump continuously running. Force–displacement curves and R–curve for DCB are on Fig. 6 .

The 5% offset evaluation was not employed for the adhesive as it yielded overly conservative values. Fracture energy was determined from the R-curve in the region where crack propagation was initiated. It is evident that there are distinct differences in further crack propagation and fracture energy between the batch left in the vacuum pump and the others. This discrepancy warrants further investigation.

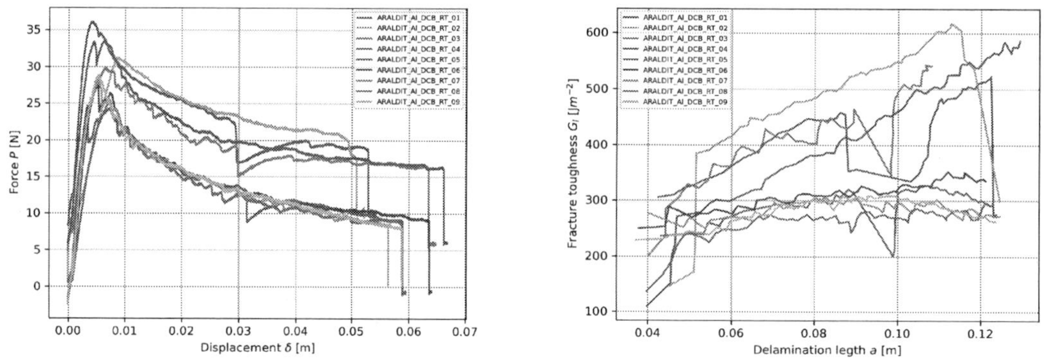

(a) Force–displacement curves (b) R–curves

Fig. 6: Results of DCB experiments.

2.5 ENF Experiments

ENF experiments were performed according to standard ASTM D7905. The loading of the specimen was carried out with a constant crosshead rate $v_{\text{ENF}} = 0.5$ mm/min. Crack length was measured as in DCB from synchronized photographs. For further evaluation, adhesive fracture toughness for mode II was computed as

$$G_{II} = \frac{3mP_{\max}^2 a_0^2}{2B}, \tag{3}$$

where $P_{\max}$ is applied load during crack propagation with initial crack length a_0. δ is load point displacement, B is width of specimen and m is term of compliance calibration coefficients. Determined using a linear least squares regression analysis of the compliance C versus crack length cubed a^3.

Force–displacement curves are shown in Fig. 7 (a) and the R–curve for ENF, which represents Mode II interlaminar fracture toughness on delamination length on Fig. 7 (b). The black × marks on the Force–displacement curves indicates the recorded crack length.

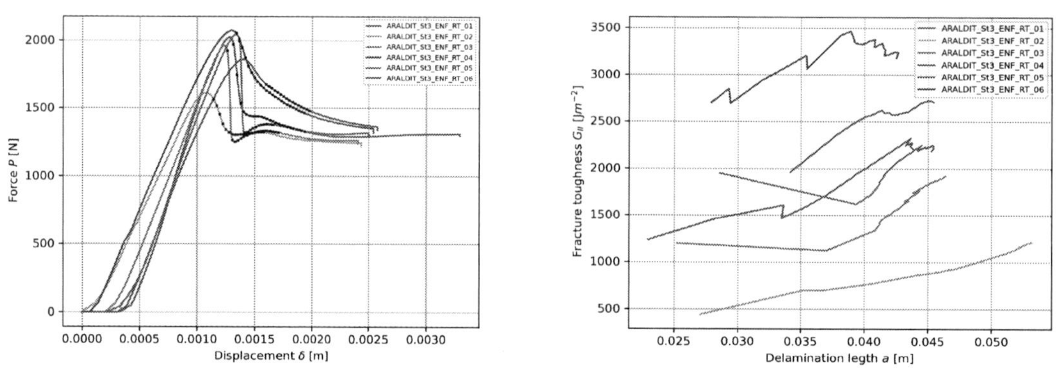

(a) Force–displacement curves (b) R–curves

Fig. 7: Results of ENF experiments.

The significant variation in results may be attributed to the thicker adhesive layer, which, when compressed during the curing process, formed rounded ends that propagated unexpectedly. This might be one of the causes. Based on these findings, it is recommended to reduce the crosshead rate when testing stiffer and thicker materials. Additionally, during fabrication, it is crucial to verify that the insert is positioned correctly and that the adhesive is not flowing around it to ensure accurate results.

3 Results

From the above experiments, material parameters were evaluated with values summarized in Table 2. The standard deviation was computed for the sample and not the whole population. CV in the table is an abbreviation for the coefficient of variation. Shear modulus was computed from Young's modulus and Poisson's constant.

Tab. 2: Evaluated parameters for complex numerical simulation.

Property	Symbol	Unit	Average	Standard deviation	CV [%]
Young's modulus	E	[GPa]	4.17	0.25	6.00
Poisson's ratio	ν	[-]	0.44	0.03	6.82
Shear modulus	G^*	[GPa]	1.45	–	–
Tensile failure strength	σ_f^T	[MPa]	19.91	1.43	7.18
Tensile failure strain	ϵ_f^T	[%]	0.51	0.05	9.80
Compression failure strength	σ_f^C	[MPa]	90.42	5.37	5.94
Shear failure strength	τ_f	[MPa]	16.67	1.91	11.46
Strain energy release rate for mod I	G_I	[Jm^{-2}]	292.78	52.72	18.01
Strain energy release rate for mod II	G_{II}	[Jm^{-2}]	1704.2	850.95	49.93

4 Conclusion

This work summarizes the methodology for obtaining complex mechanical parameters for brittle adhesives and measures these parameters for adhesive Araldit AV 138+HY 998.

Several experiments were conducted for the elastic zone parameters, including Young's modulus in tension, Poisson's constant, and Young's modulus in shear, including tensile tests, extensometer measurements, and Arcan tests. The evaluation of these parameters exhibited a low coefficient of variation, confirming the validity of the testing methodology.

The fracture initiation parameters, including tensile stress limit, compressive stress limit, shear stress limit, and ultimate tensile deformation, were assessed through tensile tests, compression tests, and Arcan tests. The tensile test revealed sensitivity to the manufacturing process of the adhesive plate. When air bubbles were present in the mixture, the tensile failure strength decreased by 35% compared to samples where the fracture occurred in areas without air bubbles.

To evaluate crack propagation parameters such as the strain energy release rate for modes I and II, specimens were prepared for Double Cantilever Beam (DCB) and End-Notched Flexure (ENF) tests. The DCB test was successfully validated for methodology but revealed uncertainty in the handling of the vacuum pump. The ENF test highlighted weaknesses in the current adhesive testing methodology.

Acknowledgement

This research work was supported by the project SGS-2022-008.

References

[1] A.M.G. Pinto et al., Shear Modulus and Strength of an Acrylic Adhesive by the Notched Plate Shear Method (Arcan) and the Thick Adherend Shear Test (TAST), Materials Science Forum 636-637 (2010) 787–792, doi: 10.4028/www.scientific.net/MSF.636-637.787.

[2] M.D. Banea, L.F.M da Silva, Adhesively bonded joints in composite materials: An overview, Proceedings of the Institution of Mechanical Engineers, Part L: Journal of Materials: Design and Applications 223 (2009) 1–18, doi: 10.1243/14644207JMDA219.

[3] G. Dean, B. Duncan, Preparation and Testing fo Bulk Specimens of Adhesives, Teddington, Middlesex, National Physical Laboratory, 1998, ISSN 1368-6550.

[4] R. Kotner, J. Heczko, J. Krystek, Validation of identified material parameters of rubber using an Arcan shear test, Materials Today: Proceedings 12(2) (2019) 404–410, doi: 10.1016/j.matpr.2019.03.142.

[5] E. M. Petrie, Handbook of Adhesives and Sealants, New York, USA, MCGraw-Hill, 2000.

Determination of the Dynamic Bending and Torsional Stiffness of a Beam using Enhanced FRF

R. Huňady [1,a], P. Palička [1], E. Dubňanská [1], M. Hančinová [1]

[1] *Technical University of Košice, Faculty of Mechanical Engineering, Letná 1/9, Košice, Slovakia*
[a] *robert.hunady@tuke.sk*

Abstract: This paper deals with the estimation of dynamic stiffness coefficients from the so-called enhanced frequency response functions (EFRFs), which are extracted from the FRF matrix obtained in the modal test. To determine the bending and torsional stiffness parameters, the synthesized EFRF obtained by the curve-fitting method is used. The given procedure is presented on the example of a free straight beam with a constant cross section.

Keywords: bending stiffness; torsional stiffness; estimation; modal analysis; EFRF.

1 Introduction

Determining the stiffness of components is a common issue in engineering practice. Most often, the tensile, flexural and torsional stiffness coefficients are determined as a structural response to static loads. For simple shaped components such as beams, shafts, etc., it is possible to find analytical formulas in the literature for the calculation of stiffness coefficients for various boundary conditions [1]. In the case of parts with complex geometry, they are determined using numerical simulations or experimental measurements. A typical practical application is vehicle body stiffness analysis to determine bending, torsional and lateral stiffness [2-5]. It is well known that these parameters have a significant influence on the comfort and driving characteristics of the car. Other examples could be found in civil engineering [6], or in the aerospace industry [7, 8]. Unlike static stiffness, dynamic stiffness exhibits different values depending on the rate of object deformation or the frequency of vibration. The dynamic stiffness coefficient is defined as the frequency dependant ratio between the amplitude of the dynamic force and the amplitude of the dynamic displacement. Based on this, it is evident that an increase in dynamic stiffness will reduce the vibration response of the system and vice versa. The static data is contained in the dynamic data (at a frequency of 0 Hz). In the following section, it will be shown how an estimate of the stiffness coefficients can be obtained from the frequency characteristics of the structure.

To determine the dynamic stiffness coefficients, we can use the so-called Enhanced Frequency Response Functions (EFRF), which are the result of transforming the responses from physical to modal space. In this way, it is possible to isolate a selected mode and express it in terms of the response of a system with one degree of freedom. The enhanced frequency response function of an arbitrary r-th mode can be determined as follows [9]

$$EFRF(\omega)_r = \{U(\omega_r)\}^H \, [H(\omega)] \, \{V(\omega_r)\} \, , \tag{1}$$

where the mode shape vector $\{U(\omega_r)\}$ and the vector of modal participation factors $\{V(\omega_r)\}$ of the r-th mode serve as discrete modal filters that amplify the response of the r-th mode, while the responses of the other modes are attenuated. Vectors $\{U(\omega_r)\}$ and $\{V(\omega_r)\}$ are obtained by singular value decomposition (SVD) of the FRF matrix $[H(\omega)]$ done for each spectral line

$$[H(\omega)] = [U(\omega)]^H \, [S(\omega)][V(\omega)] \, , \tag{2}$$

where $[S(\omega)]$ is a diagonal singular value matrix. The plot of singular values corresponds to the Complex Mode Indicator Function (CMIF)

$$CMIF(\omega)_k = S_k(\omega)\,. \tag{3}$$

If we extract the EFRF functions for the first bending and first torsional modes in this way, then the dynamic stiffness coefficient can be determined from the compliance line $\kappa(\omega)$ tangent to the left part of the EFRF curve (see Fig. 1). The tangent to the right part expresses the modal mass of the mode. Depending on the type of FRF, the compliance of the mode is given by

$$\kappa(\omega) = \frac{1}{k} \text{ - for receptance,} \quad \kappa(\omega) = \frac{i\omega}{k} \text{ - for mobility,} \quad \kappa(\omega) = \frac{-\omega^2}{k} \text{ - for inertance.} \tag{4}$$

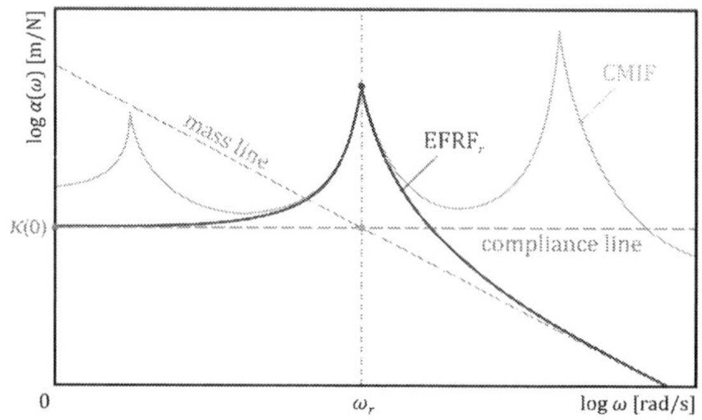

Fig. 1: Isolated response of the selected mode in the form of a receptance.

The detailed procedure for deriving the compliance curve and calculating the stiffness coefficients is given in Chapter 3.

2 Experimental Modal Analysis a Free-free Beam

The object of the modal test was a steel beam of rectangular cross-section with dimensions 40x8x400 mm. The beam was loaded quasi-freely using two rubber bands (Fig. 2a). A B&K impact hammer Type 8206 was used for excitation. The responses were measured using a Polytec PDV100 laser vibrometer. The geometric model (Fig. 2b) consisted of 18 input DOFs and 1 output DOF. The measurements were evaluated in the Pulse Reflex software. In the frequency range up to 6000 Hz, 9 modes were identified. The first bending and the first torsional mode shape of the beam with the modal parameters are shown in Fig. 3.

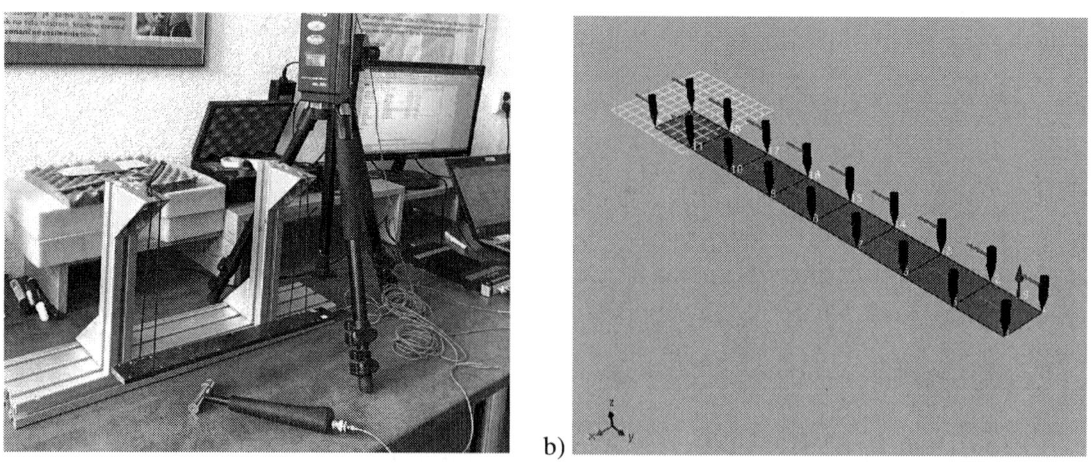

a) b)

Fig. 2: a) measurement of modal parameters, b) geometric model.

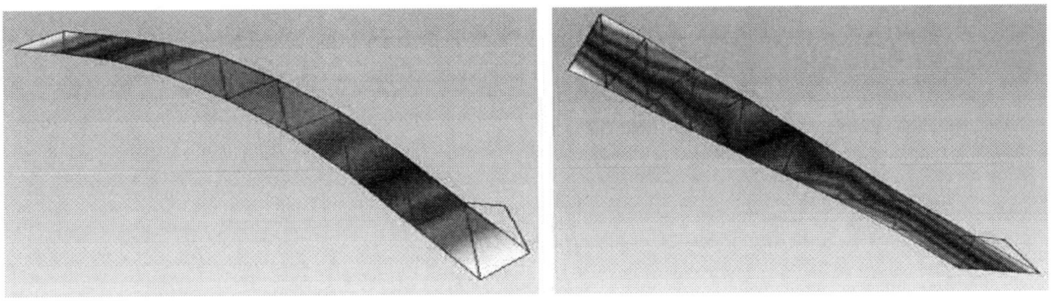

Fig. 3: The 1. bending (f_1 = 266.5 Hz, ζ_1 = 0.00386) and the 1. torsional mode (f_4 = 1496.8 Hz, ζ_4 = 0.00163).

3 Estimation of Stiffness Coefficients

All frequency response functions (FRFs) obtained from the measurements were exported to MATLAB for further processing. In the first step, they have been converted from mobility $Y(\omega)$ to receptance $\alpha(\omega)$ according to the well-known relation

$$\alpha(\omega) = \frac{Y(\omega)}{\mathrm{i}\omega}. \tag{5}$$

Subsequently, the FRF matrix was constructed and based on the procedure described in Chapter 1, the CMIF and EFRF functions were extracted (see Fig. 4). As can be seen in the figure, the EFRF curves do not exactly correspond to the ideal response of a SDOF system. This shortcoming is due to the limited number of data in the eigenvectors that are used as modal filters. Nevertheless, the frequency response in the vicinity of the natural frequency is sufficiently isolated from the other modes and can be used to construct a synthesized EFRF.

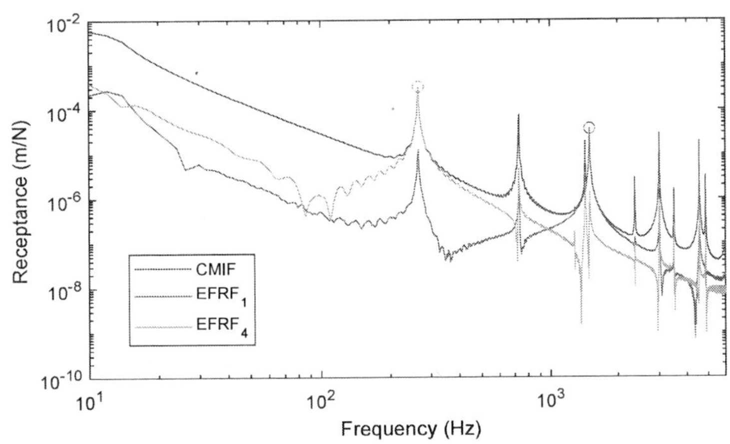

Fig. 4: CMIF and EFRF plot of the first bending and the first torsional mode.

Since the response of an arbitrary mode is most dominant around its resonant frequency, the so-called spectral bell was first defined. This bell is represented by a small region of the resonant peak. Basically, it consists of a small number of data, as can be seen in Fig. 5. The synthesized EFRF was determined using the curve fitting method, starting from the equation

$$EFRF(\omega)_{synt,r} = \frac{Q_r}{\lambda_r - \omega^2}, \tag{6}$$

where Q_r is the modal scale factor, $\lambda_r = \Omega_r^2\left(1 + \mathrm{i}\frac{\zeta_r}{2}\right)$ is the complex pole, Ω_r is the natural angular frequency, and ζ_r is the damping of the r-th mode.

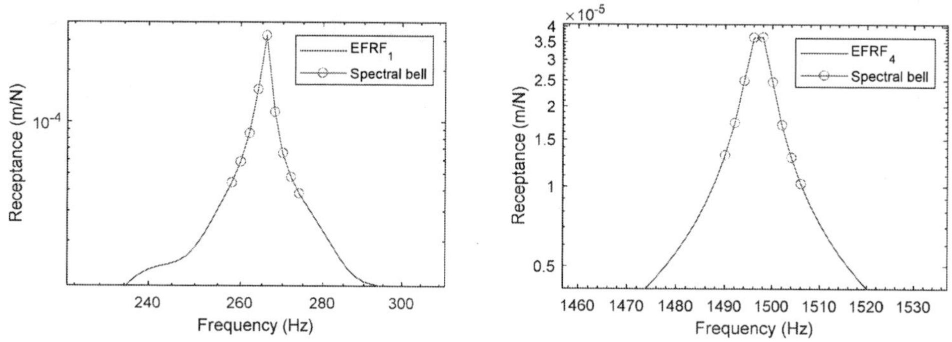

Fig. 5: Spectral bell defined on EFRF curve.

MATLAB optimization tools were used to determine the three unknown parameters Q_r, Ω_r, and ζ_r. The objective function was to minimize the differences between the measured and synthesized EFRF on the frequency band ω_{sb} of the spectral bell of the corresponding mode

$$\min\left(\left\|EFRF(\omega_{sb})_{exp,r} - EFRF(\omega_{sb})_{synt,r}\right\|\right) . \tag{7}$$

The modal parameters (f_r, ζ_r) extracted from the measurements were used as initial values of the design variables. The initial estimate of the modal scale factor was determined as follows

$$Q_r = 2\, EFRF(\Omega_r)\, \Omega_r^2\, \zeta_r, \tag{8}$$

where $\Omega_r = 2\pi f_r$. The `fminsearch` function, which searches for the minimum of an unconstrained multivariate function using a derivative-free method, was used for the computation. The optimized values of the search variables are shown in Tab. 1. Fig. 6 and Fig. 7 show the resultant synthesized EFRF plot of the first bending and the first torsional mode, respectively.

Tab. 1: Modal parameters obtained in the curve-fitting process.

Mode	Q_1 (Nm^{-1}s^{-2})	f_1 (Hz)	ζ_1 (–)	Mode	Q_4 (Nm^{-1}s^{-2})	f_4 (Hz)	ζ_4 (–)
bending	6.272669	265.66	0.003265	torsional	11.356699	1496.98	0.001622

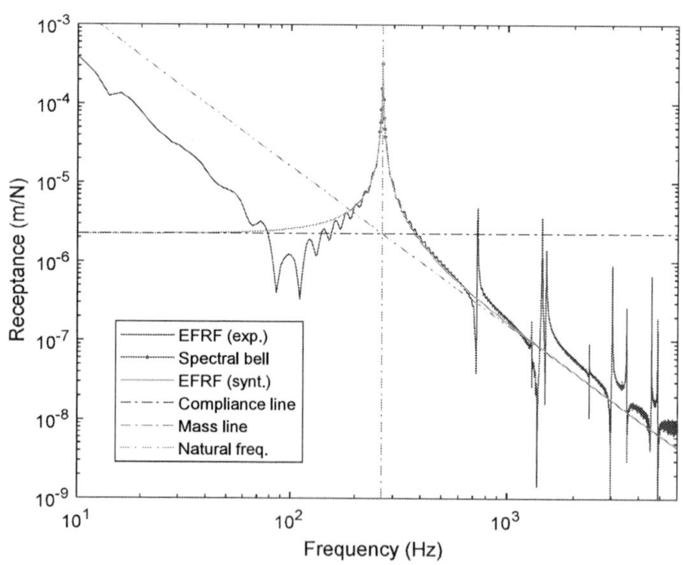

Fig. 6: EFRF of the first bending mode.

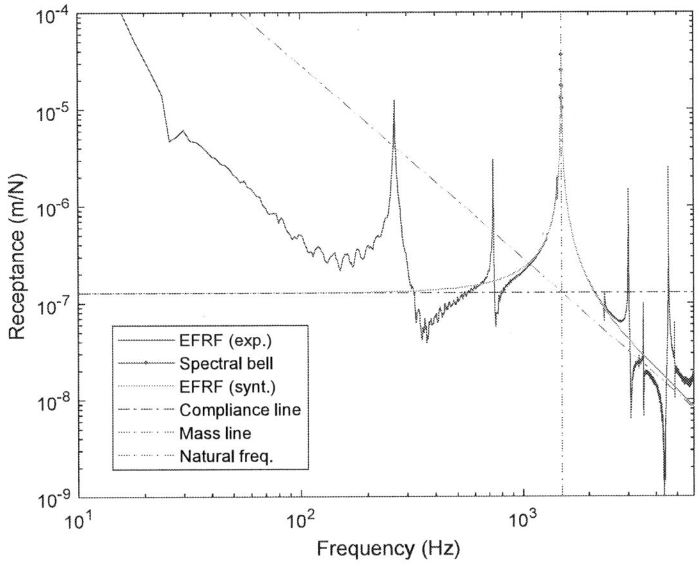

Fig. 7: EFRF of the first torsional mode.

Since the spectra shown in Fig. 6 and Fig. 7 represent the receptance, i.e. the mechanical compliance (m/N), the stiffness is given as the inverse value. For a free-free beam, the bending stiffness coefficient k_b is determined as

$$k_b = \frac{1}{2\kappa(0)}, \tag{9}$$

where $\kappa(0)$ is the compliance value at zero frequency taken from the synthesized EFRF. To calculate the torsional stiffness coefficient k_t of the beam, it is necessary to use the equation

$$k_t = \frac{b^2}{4\kappa(0)}, \tag{10}$$

where b is the beam width. Equation (10) converts the value of the stiffness coefficient from N/m to Nm/rad. The resultant coefficients are listed in Tab. 2.

Tab. 2: Dynamic bending and torsional stiffness coefficients of a free-free beam.

Mode	$\kappa(0)$	k_b (N/m)	Mode	$\kappa(0)$	k_t (Nm/rad)
1. bending	$1.9532 \cdot 10^{-6}$	255990	1. torsional	$1.19424 \cdot 10^{-7}$	3349

4 Conclusion

The aim of the paper was to present an alternative approach for determining the stiffness parameters of the structure. Using the example of a free beam, it was shown how the dynamic stiffness coefficients in bending and torsion can be determined from the frequency response functions. The key is to isolate the frequency response of the corresponding mode, for which the enhanced frequency response function can be advantageously used. As has been shown the degree of isolation depends on the eigenvectors, which means that increasing the number of DOFs of the geometric model can lead to better estimation.

Acknowledgement

This work was supported by the Science Grant Agency of the Ministry of Education, Science, Research and Sport of the Slovak Republic and the Slovak Academy of Sciences under the project VEGA 1/0324/24.

References

[1] F. Trebuňa, F. Šimčák, Resistance of Elements of Mechanical Systems, 1. ed., Technical University of Košice, Košice, 2004. (in Slovak)

[2] L. Shengqin, F. Xinyuan, Study of structural optimization design on a certain vehicle body-in-white based on static performance and modal analysis, Mechanical Systems and Signal Processing 135 (2020) 106405, https://doi.org/10.1016/j.ymssp.2019.106405.

[3] D. Mundo, R. Hadjit, S. Donders, M. Brughmans et al., Simplified modelling of joints and beam-like structures for BIW optimization in a concept phase of the vehicle design process, Finite Elements in Analysis and Design 45(6) (2009) 456–46, https://doi.org/10.1016/j.finel.2008.12.003.

[4] A. Ramachandran, H. Reddy, T. Chavali, A. Hukar et al., Body-in-White Joint Stiffness Sensitivity Analysis, SAE Technical Paper (2019) 2019-01-5012, https://doi.org/10.4271/2019-01-5012.

[5] S.Y.R. Aizzat, R. Rahizar, M.H. Sallehuddin, A. Anuar, Improving the Dynamic Characteristics of Body-in-White Structure Using Structural Optimization, The Scientific World Journal 2014 (2014) 190214, https://doi.org/10.1155/2014/190214.

[6] L. Xiang, T. Shitan, Z. Xueyi, L. Xiao et al., Dynamic response analysis for bridges subjected to moving vehicle loads by using the analytical dynamic stiffness method, Computers & Structures 292 (2024) 107240, https://doi.org/10.1016/j.compstruc.2023.107240.

[7] S. Qiao, J. Jiao, Y. Ni, H. Chen, X. Liu, Effect of Stiffness on Flutter of Composite Wings with High Aspect Ratio, Mathematical Problems in Engineering 2021 (2021) 6683032, https://doi.org/10.1155/2021/6683032.

[8] Q. Ming, Z. Wenguo, L. Shu, The Effect of Torsional and Bending Stiffness on the Aerodynamic Performance of Flapping Wing. Aerospace 10(12) (2023) 1035, https://doi.org/10.3390/aerospace10121035.

[9] R.J. Allemang, D.L. Brown, A complete review of the complex mode indicator function (CMIF) with applications, in proc.: International Conference on Noise and Vibration Engineering (ISMA) (2006), Katholieke Universiteit Leuven, Belgium, 3209–3246.

Experimental Study of Fundamental Mechanical Properties of Ultra-Thin 3D-Printed Titanium Alloy with Beta Structure

A. Jíra[1,*], Z. Tolde[2], J. Fojt[3]

[1] *Czech Technical University in Prague, Faculty of Civil Engineering, Department of Mechanics, Czech Republic*

[2] *Czech Technical University in Prague, Faculty of Mechanical Engineering, Department of Physics, Czech Republic*

[3] *University of Chemistry and Technology Prague, Faculty of Chemical Technology, Department of Metals and Corrosion Engineering, Czech Republic*

* *jira@fsv.cvut.cz*

Abstract: 3D printing of beta-titanium alloys for biomedical applications is currently in high demand, both for material reasons and for the ability to produce highly complex replacements, often directly tailored to the patient. These individual implants require a porous surface that is printed from thin parts at the limits of the 3D printer's capabilities. The thin structures are then very susceptible to changes in mechanical properties due to poor print quality, and building a corresponding numerical model is very complicated. The aim of this work is to experimentally describe the basic mechanical properties of thin 3D-printed titanium alloy samples with a beta structure. These values can then serve as input quantities for numerical modeling via particle models.

Keywords: 3D printing; surface etching; titanium alloy; porosity; mechanical testing.

1 Introduction

In the past decade, there has been a revolutionary transformation in manufacturing due to 3D printing, enabling the creation of intricate structures with high precision. In biomedicine, this technology has become pivotal for the development of implants and joint replacements that can better match individual anatomical characteristics of patients [1, 2]. The key advantage of using 3D printing with these materials lies in the utilization of complex porous structures, such as gyroid-based designs [3], which ensure optimal integration of the implant with surrounding bone tissue. However, these structures are often very thin (typically 150 – 600 μm), pushing the limits of current printers and the powders used.

A critical factor for a successful 3D-printed implant is the material used and its biomechanical reliability. Most current replacements are made from the titanium alloy Ti6Al4V, which is well-proven and has been used for a long time. However, it also has disadvantages, such as a high Young's modulus and potential toxicity, making it worth considering alternatives. Possible replacements include beta-structured titanium alloys, such as Ti25Nb4Ta8Sn [4, 5, 6]. The Ti25Nb4Ta8Sn alloy is bioactive, non-cytotoxic, and serves as a suitable alternative to the widely used Ti6Al4V alloy in terms of mechanical properties and corrosion resistance [7]

In real-world applications, experimental measurements often do not correspond with numerical models due to internal material porosity and incomplete fusion of powder particles at the structure's edges [8].

This work serves as a pilot study on the behavior of thin 3D printed samples, aiming to experimentally assess their mechanical properties. The values obtained will form the basis for creating a numerical model. Numerical modeling and simulations are crucial for precise design and optimization of these structures based on desired mechanical properties.

2 Methodology

Sample preparation was focused on 3D printing (type Dog bone) specimens for uniaxial tensile testing to obtain fundamental material characteristics for subsequent numerical modeling. For this purpose, both conventional titanium alloy Ti6Al4V and an experimental alloy Ti25Nb4Ta8Sn were utilized. The Ti6Al4V specimens underwent surface etching, while the Ti25Nb4Ta8Sn specimens were evaluated in different groups:

- Group I: Ti25Nb4Ta8Sn samples without supplementary post-processing treatments.
- Group II: Ti25Nb4Ta8Sn samples subjected to post-annealing to alleviate internal stresses.
- Group III: Ti6Al4V specimens without additional post-processing treatments.
- Group IV: Ti6Al4V samples subjected to surface etching and removal of imperfectly fused powder grains through washing.

These experimental groups were essential for evaluating and comparing the mechanical properties and structural integrity of the printed specimens under different conditions.

2.1 Preparation of samples

Groups III and IV samples were produced using the SLM technique from Ti6Al4V titanium alloy (Rematitan CL) with a maximum grain size of 63 μm. These samples were fabricated in collaboration with ProSpon, Ltd. (Kladno, Czech Republic) using an M2 Cusing machine. The 3D printing was conducted in an argon environment with 0.5% oxygen maintained within the welding chamber. After printing, the samples underwent thermal processing under vacuum to relieve internal stresses, involving gradual heating to 840 °C over 4 hours, holding at 840 °C for 2 hours, and then cooling to room temperature. Additionally, samples from group IV were subjected to chemical etching to remove imperfectly fused powder [9] particles.

For 3D printing samples from groups I and II, Ti25Nb4Ta8Sn alloy powder with a particle size range of 10-120 μm and an average particle size of 50 μm, supplied by Advanced Metal Powders s.r.o., was used. The selective laser melting (SLM) method was employed for 3D printing the beta-titanium alloy in collaboration with the Institute of Physics of the Czech Academy of Sciences, followed by thermal treatment in cooperation with UJP Prague. The samples were annealed at 1000 °C for 1 hour and then gradually cooled in the furnace.

2.2 Mechanical testing in uniaxial tension

Mechanical testing was conducted using a LiTeM machine (Italy). Dog bone-shaped specimens (Fig. 1) from groups I-IV were produced with thicknesses of 500, 750, and 1000 μm and subjected to uniaxial tensile testing. The test parameters were chosen according to EN 10993 standards, considering the potential future use for implants. The test was performed in controlled displacement mode at a rate of 0.04 mm/min.

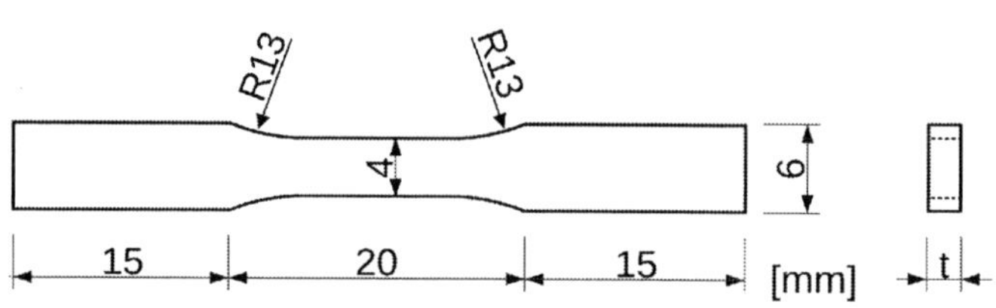

Fig. 1: The geometry of dog bone-shaped test specimens for uniaxial tensile testing features a thickness (t) determined by the printing capabilities, set in increments of 250 μm.

Results

As a basic comparison to illustrate the development of mechanical properties of thin samples, uniaxial tensile testing was performed. The graphs in Fig. 2 show representative force-displacement curves for different sample thicknesses (t). The Ti25Nb4Ta8Sn alloy samples were tested on five specimens without subsequent processing (blue line) and on five specimens that underwent annealing to remove internal stresses (red line). The Ti6Al4V alloy samples were tested on five specimens without additional treatment (blue line) and on five specimens after etching to remove imperfectly welded powder particles (green line).

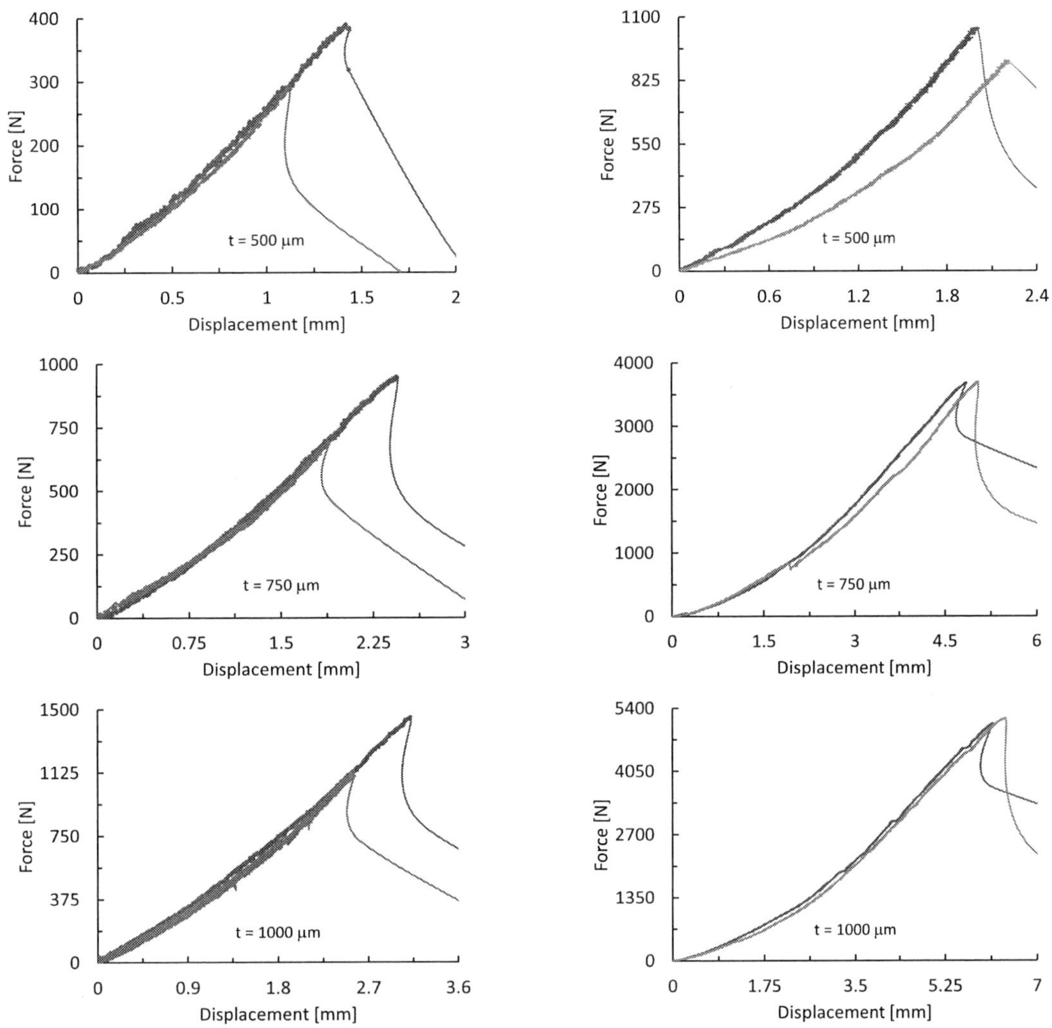

Fig. 2: Graphs depicting the force-displacement relationship from uniaxial tensile testing for materials Ti25Nb4Ta8Sn (left column) and Ti6Al4V (right column). Ti25Nb4Ta8Sn without subsequent processing – blue line, Ti25Nb4Ta8Sn with annealing – red line. Ti6Al4V without additional treatment – blue line, Ti6Al4V after etching – green line.

The standard Ti6Al4V alloy exhibits significantly higher tensile strength compared to the Ti25Nb4Ta8Sn alloy (Tab. 1). This difference can be attributed to the significantly higher porosity in the beta-structured alloy. This high porosity also leads to a substantial reduction in strength after annealing, although there is no significant change in Young's modulus. In contrast, the conventional alpha-beta alloy did not show a decrease in tensile strength in the etched samples, but a significant decrease in Young's modulus was observed in the thinnest samples based on the graphs. As thickness increases, the differences in the force-displacement curves diminish, indicating that the quality of 3D printing plays a less decisive role.

The failure of the samples during uniaxial tensile testing mainly occurred along the grain boundaries. Additionally, a significant number of imperfectly melted powder particles were observed, resulting in very low tensile strength and fracture forces (Fig. 3). This phenomenon was only observed in the Ti25Nb4Ta8Sn alloy, which the authors attribute to incorrect printing settings. Optimization of printing parameters will be the subject of further investigation and is not part of this study. However, for the purposes of numerical modeling and particle model development, this high porosity is advantageous as it allows for more detailed particle modeling.

The values of the Young's modulus can be determined from the graphs in Fig. 2. The average values range from 3600 to 3900 MPa for the Ti25Nb4Ta8Sn alloy (for thicknesses of 500 to 1000 μm). In comparison, the modulus of elasticity for the Ti6Al4V alloy is in the range of 7000 to 8000 MPa. Post-processing modifications

t [μm]		Ti25Nb4Ta8Sn with annealing	Ti25Nb4Ta8Sn without annealing	Ti6Al4V with etching	Ti6Al4V without etching
500	F_{max} [N]	407.5 ± 11.8	207.7 ± 29.6	1009.4 ± 43.5	760.0 ± 85.4
	σ_{max} [MPa]	203.8 ± 5.9	135.3 ± 14.8	504.7 ± 21.8	380.3 ± 75.0
750	F_{max} [N]	889.6 ± 38.6	674.0 ± 86.1	3570.4 ± 129.5	3571.1 ± 105.8
	σ_{max} [MPa]	296.5 ± 12.9	224.7 ± 28.7	1190.1 ± 83.2	1190.4 ± 61.9
1000	F_{max} [N]	1442.6 ± 69.9	1010.9 ± 101.0	4579.4 ± 177.1	4958.9 ± 133.0
	σ_{max} [MPa]	360.6 ± 16.7	252.7 ± 25.3	$1144.9.7 \pm 144.9$	1246.5 ± 45.7
		Group I	Group II	Group III	Group IV

Tab. 1: Average values of yield strength and ultimate tensile strength for different types of samples.

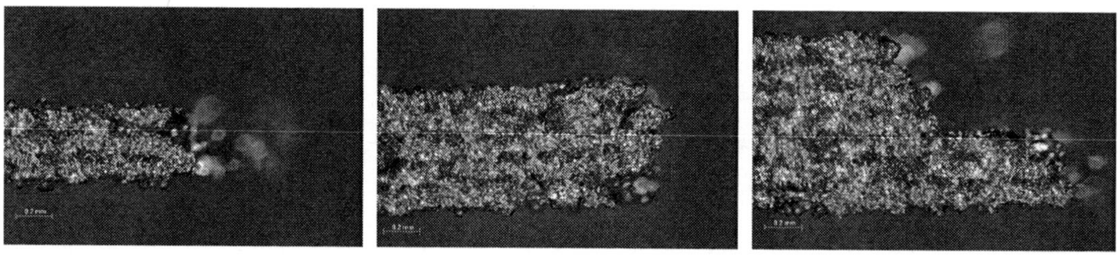

Fig. 3: Samples from group I with thicknesses of 500, 750, and 1000 μm showed visible internal structures due to imperfect 3D printing.

have a negligible effect on the resulting modulus of elasticity.

3 Conclusions

Uniaxial tensile testing was conducted as a fundamental comparison to illustrate the development of mechanical properties in thin samples. The standard Ti6Al4V alloy exhibits significantly higher ultimate tensile strength values compared to the Ti25Nb4Ta8Sn alloy. This difference can be attributed to the significantly higher porosity in the beta-structured alloy. This high porosity ($\sim$16–19 %) also leads to a substantial reduction in strength after annealing, although there is not a notable change in the Young's modulus. Conversely, for the conventional alpha-beta alloy, there was no decrease in ultimate tensile strength for etched samples, but significant decreases in Young's modulus were observed for the thinnest samples. As the thickness increases, the differences in the force-displacement curves diminish, suggesting that the quality of 3D printing plays a less critical role.

Acknowledgement

We are thankful for financial support provided by The Czech Science Foundation (GACR) grant No. 23-04971S and Czech Technical University in Prague grant SGS No. SGS23/152/OHK1/3T/11.

References

[1] D. Serrano, et al., 3D Printing Technologies in Personalized Medicine, Nanomedicines, and Biopharmaceuticals, Pharmaceutics. vol. 15 (2023).

[2] B. Li, et al. Application and Development of Modern 3D Printing Technology in the Field of Orthopedics, BioMed Research International. vol. 2022 (2022) 1-15.

[3] A. Jíra, et al., Variability of the Gyroid Structure for Biomedical Applications, 59th International Scientific Conference on Experimental Stress Analysis, Litomyšl, Czech Republic, 2021.

[4] Z. Tolde, V. Starý, L. Cvrček, M. Vandrovcová, J. Remsa, S. Daniš, Growth of a TiNb adhesion interlayer for bioactive coatings, Materials Science and Engineering: C. vol. 80 (2017). https://doi.org/10.1016/j.msec.2017.07.013.

[5] M. Vandrovcova, Z. Tolde, P. Vanek, V. Nehasil, M. Doubková, M. Trávníčková, J. Drahokoupil, E. Buixaderas, F. Borodavka, J. Novakova, L. Bacakova, Beta-Titanium Alloy Covered by Ferroelectric Coating–Physicochemical Properties and Human Osteoblast-Like Cell Response, Coatings. vol. 11 (2021). https://doi.org/10.3390/coatings11020210.

[6] I. Jirka, M. Vandrovcová, O. Frank, Z. Tolde, J. Plšek, T. Luxbacher, L. Bačáková, V Starý, On the role of Nb-related sites of an oxidized β-TiNb alloy surface in its interaction with osteoblast-like MG-63 cells, Materials Science and Engineering: C. vol. 33 (2013) 1636-1645. https://doi.org/10.1016/j.msec.2012.12.073.

[7] V. Hybasek, J. Fojt, J. Malek, E. Jablonska, E. Pruchova, L. Joska, T. Ruml, Mechanical properties, corrosion behaviour and biocompatibility of TiNbTaSn for dentistry, Materials Research Express. vol. 7 (2020) 13. https://doi.org/10.1088/2053-1591/ab62f7.

[8] A. Jíra, et al., Mechanical Properties of Porous Structures for Dental Implants: Experimental Study and Computational Homogenization, Materials. vol. 14 (2021).

[9] J. Fojt, Z. Kacenka, E. Jablonska, V. Hybasek, E. Pruchova, Influence of the surface etching on the corrosion behaviour of a three-dimensional printed Ti–6Al–4V alloy, Materials And Corrosion. 71 (2020) 1691-1696. https://doi.org/10.1002/maco.202011658

[10] EN ISO 10993 standards – Testing of Medical Devices.

Influence of Temperature Change on Young Modulus of Elasticity of CFRP

B. Kropík[1,*], M. Matušů[1], Z. Padovec[1], T. Mareš[1]

[1] *Department of Mechanics, Biomechanics and Mechatronics, Faculty of Mechanical Engineering, Czech Technical University in Prague, Technická 4, 160 00, Prague, Czech Republic*

* *bohumil.kropik@fs.cvut.cz*

Abstract: Repeated tensile testing at low stress level was done at different temperature levels to estimate the temperature influence on the Young modulus of the composite material. Testing was done for temperatures higher than the room temperature and also for lower temperatures with use of liquid nitrogen cooling. Tests were held for two different types of fibres and the results indicate change in the Young modulus depending on the temperature.

1 Introduction

Composite materials, which in this context refer to a Carbon Fibre Reinforced Polymers (CFRP), are often designed to enhance the mechanical strength of structures with the aim of reducing the assembly weight or increasing structural potential. Operating temperatures often do not differ significantly from room temperature, but in some cases they may vary considerably. Some information on temperature-related behaviour regarding the mechanical properties of CFRP can be found in [1] for temperatures higher than room temperature and in [2] for temperatures lower than 0 °C. Therefore, for our design purposes, it was decided to measure the relationship between the Young modulus of elasticity and temperature. The CFRP used in this study comprised two types of carbon fibres: High Strength (HS), partially documented in [3], and Ultra-High Modulus (UHM) fibres. These specimens were subjected to repeated measurements during the elastic phase of the tensile test to determine whether the temperature variations affect the stiffness of the composite material and to what extent, particularly in our case.

2 Specimens

The composite used for the specimens was made using a filament winding process on a square cross-section mandrel. Of this, flat planes of about 1000x120x5 mm were cut and delivered by the manufacturer. Specimens were manually cut to an approximate size of 200x40x5 mm, representatives are shown in Fig. 1. In total, 10 specimens were made using High Strength (HS) fibres and 8 using Ultra High Modulus (UHM) fibres, both with the epoxy resin and almost identical lay-up. In the centre of each specimen, 1 linear Strain Gauge (SG) was installed on each side orientated in the direction of loading.

3 Loading Scenario

Specimens were loaded during the tensile test on the TIRA 2300 testing machine with a displacement of 1 $\frac{\text{mm}}{\text{min}}$. With the idea of using the specimens repeatedly at different temperature levels, the loading was limited to not damage the specimens. After the installation of the specimen on the loading machine, three loading cycles between 1 and 10 kN were performed at each temperature level. Measurements were first performed at 20 °C. Using a thermal chamber, we continued at 60 °C and then at 100 °C. The next step was the identification of the behaviour at lower temperatures. The specimens were cooled with the use of liquid nitrogen in our thermal chamber, at least 30 minutes before loading. The plan was to measure behaviour at 0 °C, -20 °C and -40 °C.

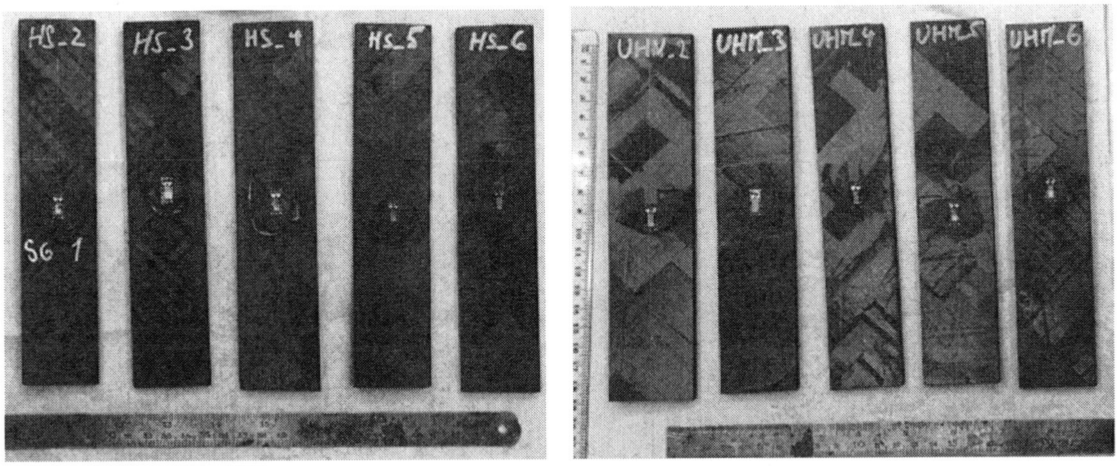

(a) representatives of HS specimens (b) representatives of UHM specimens

Fig. 1: Specimens.

The SGs installed on the specimens were connected in a quarter-bridge connection, so only the tensile strain ε_T was evaluated using Eq. (1) assuming the ε_1 and ε_2 being the signals from each SG. From each cycle, the linear section between 4 and 5.5 kN was used for the evaluation to neglect any clearances of the loading setup.

$$\varepsilon_T = \frac{\varepsilon_1 + \varepsilon_2}{2} \tag{1}$$

4 Results

The results are presented separately for the elevated and decreased temperatures. Young modulus is presented related to the temperature level of 20 °C for each material. In Fig. 2 the results for the HS specimens at elevated temperature levels are presented in a typical form of a box plot. The green triangle represents the average value and the orange horizontal line shows the median. The box is sized between 1. and 3. quartiles and whiskers are up to 1.5 times the size of the box, depending the position of measured values. It could be seen that the data vary a lot between the specimens but there is a decreasing trend of stiffness of the specimens with the rising temperature level. A similar result is with the UHM specimens presented in Fig. 3 that appeared to decrease stiffness more significantly at the temperature level of 100 °C. The scatter in the measured data is attributed to the manual division of the specimens which resulted in a width variation of up to 3 mm. This difference in the design leads to the differences in the SG position according to the layup of the composite and also affects the measured signal. Other influences could be the surface of the composite material itself, which varied on each side of the specimen, considering the inner (mandrel) and outer surfaces, and the error of installation of the specimens in the loading machines, which was done manually.

As the measurements at the decreased temperature levels with liquid nitrogen cooling were done for the first time, it was decided to pick two representative specimens of each type for the measurement at 0 °C. This measurement showed that it is necessary to protect the SG from the condensed moisture, which was not done for the increased temperature levels. After SG protection, the specimens were loaded at room temperature again to inspect the behaviour. As they were considered not damaged, the temperature levels of -20 °C and -40 °C were measured with the jaws rotated by 90 °to have the outlet of liquid nitrogen from the side. The results of this measurement are not presented in more detail, as there was again a decrease in the stiffness of the specimens with decreasing temperature, which was unexpected. Repeated measurement at room temperature showed a decrease in stiffness of approximately 20 % for HS specimens and 30 % for UHM specimens after measurements at lower temperature levels. We assume that the specimens were damaged during measurement. The damage to the specimens was probably caused by two main factors. The first could be the frozen moisture from condensed air that could damage the outer layers of the composite. Specimen loaded at 0 °C is presented in Fig. 4 but considering that there was no damage after this loading, the main factor is considered

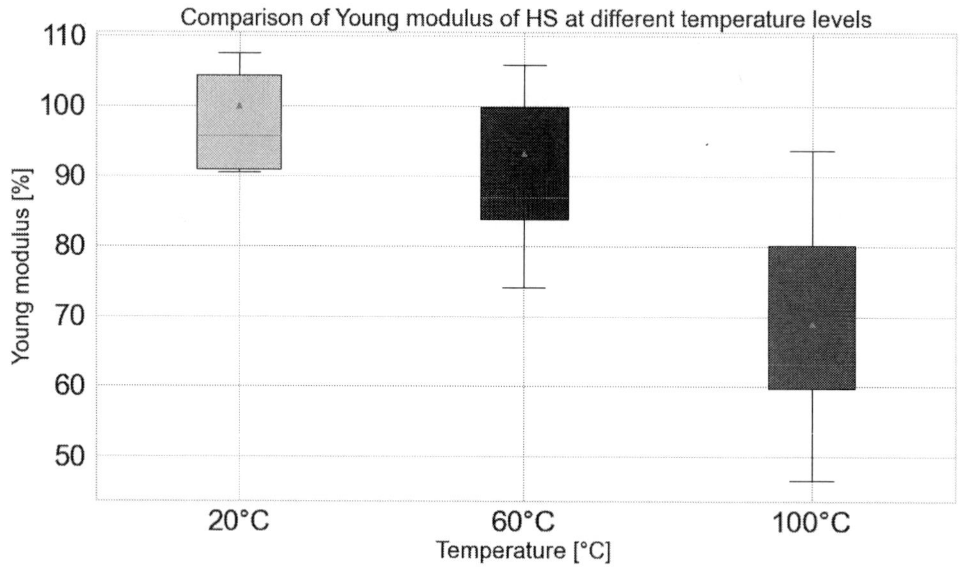

Fig. 2: HS results.

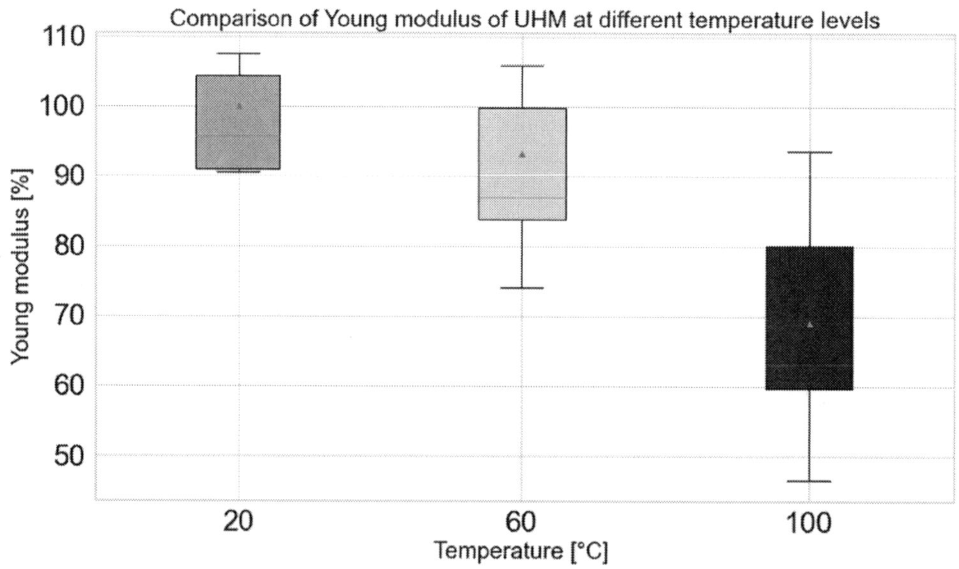

Fig. 3: UHM results.

Fig. 4: Specimen during lower temperature loading.

to be the injection of liquid nitrogen into the chamber. The chamber was not perfectly sealed, which resulted in a continuous injection of liquid nitrogen. The construction of the chamber has an outlet that aims strictly at the specimen from a close distance. The thermal chamber used was not specially designed for the tests of composite materials. As liquid nitrogen is often used successfully for the cooling of composites, we assume that the chamber needs to be perfectly sealed to minimise the amount of injected nitrogen and to control the cooling to be slower, preferably with outlet not in the direction of specimens or with some protection.

5 Conclusion

A decrease in stiffness was observed for increased temperature levels of approximately 9 % (60 °C) and 15 % (100 °C) for the HS composite and about 8 % (60 °C) and 31 % (100 °C) for the UHM composite. Although the scatter of the measured data, decrease in stiffness could be considered an expected behaviour considering the literature, but more tests, especially on more precise specimens, should be done to get more representative values. The measurement with liquid nitrogen cooling proved to be more complicated. To obtain useful results, the experimental setup and cooling, especially liquid nitrogen injection, must first be solved to avoid damaging the specimens. However, the operating conditions proved to be important to take into account during design.

Acknowledgement

This study received support from the Grant Agency of the Czech Technical University in Prague, under grant No. SGS24/123/OHK2/3T/12.

References

[1] Y. Yang et al., Study on Tensile Properties of CFRP Plates under Elevated Temperature Exposure, Materials 12 (2019), doi: 10.3390/ma12121995.

[2] Z. Jia et al., An experimental investigation of the temperature effect on the mechanics of carbon fiber reinforced polymer composites, Composites Science and Technology 154 (2017), doi: 10.1016/j.compscitech.2017.11.015.

[3] M. Matušů et al., Influence of Temperature Increase on Young Modulus of Elasticity of CFRP, in proc.: Applied Mechanics 2021: 22nd International Science Conference for Young Researchers, Book of Full Texts (2021), Technical University of Liberec, Liberec, 36–41.

Prediction of fatigue life of 42CrMo4+QT steel using self-heating effect

Martin Matušů [1,2,a], Jan Papuga [1]

[1]Department of Mechanics, Biomechanics and Mechatronics, Faculty of Mechanical Engineering, Czech Technical University in Prague, Technická 4, 160 00, Prague, Czech Republic

[2] Department of Mechanical and Environmental Engineering, OTH Amberg-Weiden, Germany

[a]martin.matusu@fs.cvut.cz

Abstract: This study proposes a novel approach that takes advantage of the self-heating effect during cyclic loading, showing promise for the prediction of fatigue life. The self-heating phenomenon offers valuable insight into material behaviour, but the stabilised temperature increase used as the key parameter might be absent in some cases. Introducing a modified version of Fargione's method, the study addresses such scenarios in the temperature response of specimens from 42CrMo4+QT steel examined here. Experimental investigations on five datasets, each with different machining parameters, analyse how they influence fatigue life and residual stresses. The findings indicate that the proposed approach could also be applied to materials lacking temperature increase. Challenges related to variable-limiting energy are explored, along with proposed solutions for future evaluation. In summary, this study introduces a promising path to assess fatigue life in materials with an unconventional temperature response during cyclic loading.

Keywords: Fargione's method, self-heating effect, 42CrMo4, fatigue life, surface quality

Nomenclature:

a_1, a_2	[K·MPa^{-1}], [-]	Coefficients for fitting the relation between σ_a and Φ
N_{1-2}	[-]	Number of cycles before transition to second phase
N_f	[-]	Number of cycles to failure
NETD	[mK]	Noise Equivalent Temperature Difference
Φ	[K]	Limiting energy
Φ_0	[K]	Limiting energy related to first phase
Φ_1	[K]	Limiting energy related to linear evolution in second phase
Φ_c	[K]	Limiting energy related to constant temperature delta in second phase
Φ_f	[K]	Limiting energy related to the last phase before break
R	[-]	Load ratio (minimum to maximum load)
R_a, R_z	[μm]	Mean arithmetic roughness, Maximum peak to valley height roughness
R_0	[K·s^{-1}]	Initial temperature rate
R_1	[K·s^{-1}]	Temperature rate in the second phase of temperature evolution
R_y	[K·s^{-1}]	Temperature rate at the last phase before break
SH		Self-Heating
σ_a	[MPa]	Stress amplitude
σ_{FL}, *FL*	[MPa]	Fatigue limit
T_{abs}	[K]	Absolute ambient temperature
T_{loaded}	[K]	Temperature of a dynamically loaded specimen
$T_{reference}$	[K]	Temperature of a reference specimen
Θ	[K]	Stabilized temperature increase

1 Introduction

Materials may fail under cyclic loading even when the load amplitude does not reach the tensile or yield strengths. The S-N (Stress-Life) curve is formed for the specific levels of the stress (S) cycles, which are repeated for some obtained numbers (N) of cycles till break. Predicting cycles to failure is crucial in engineering applications [1]. However, traditional methods for assessing experimental fatigue life are costly and time-consuming due to equipment requirements and the need for testing numerous specimens over many cycles to establish the S-N curve. Standard guidelines [2] suggest the use of 6-12 samples for research purposes and 12-24 specimens for structural component design. Reducing the number of samples while maintaining the accuracy of the S-N curve would yield significant cost and time savings.

The fatigue life prediction methods discussed here are based on measuring the surface temperatures of a specimen, which fluctuate during cyclic loading. This fluctuation is often monitored using an infrared camera. The difference between the ambient temperature and the surface temperature of the loaded sample is an outcome of cyclic loading, which correlates with intrinsic damage and the response of the structure to loading. This phenomenon is referred to as the Self-Heating (S-H) effect. This approach has demonstrated promising results in both steel specimens [3; 4] and aluminium specimens [5; 6].

The S-H phenomenon relates to the heating of a sample during cyclic loading, where stress amplitudes induce plastic deformations classified as damaging and non-damaging [7]. By mapping material response based on surface temperature, we can effectively analyse fatigue behaviour, including the determination of theoretical fatigue limits [8]. The self-heating effect is inherent in the response of the material to cyclic loading [9; 4]. During repeated loading at a constant amplitude, the temperature monitored on the specimen surface typically exhibits three distinct phases illustrated in Fig. 1, each characterised by specific parameters.

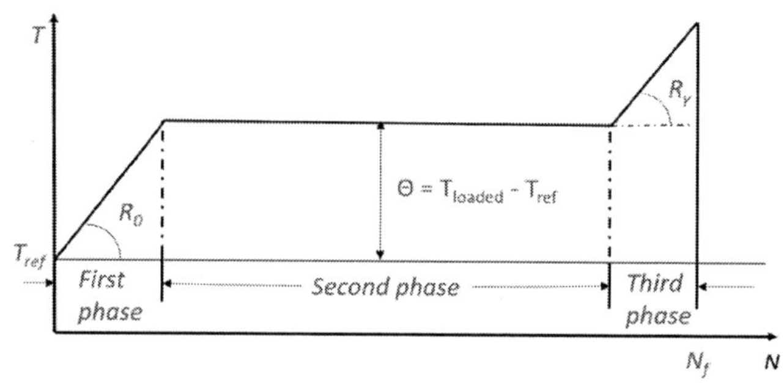

Fig. 1 Schematic depiction of three different phases in temperature response during constant-amplitude cyclic loading until failure at N_f number of cycles.

The initial phase (Phase 1) of the temperature evolution during cyclic loading at a constant load amplitude is marked by a rapid temperature increase. This phase is characterised by the variable R_0, representing the temperature derivative with respect to the number of cycles. The use of this rate to predict fatigue life was previously demonstrated in a study of reversed bending [10].

Phase 2 is of significant importance in this study, representing the constant temperature Θ response of the sample throughout most of its lifespan. This phase is characterised by a steady state, where input work in the system is equally dissipated into the surrounding environment. This parameter is crucial for many works, as seen in the references [11; 12].

The final phase is marked by a rapid increase in temperature that occurs briefly during crack growth before the sample reaches its breaking point. This phase can be effectively delineated by R_y rate, representing the temperature derivative with respect to the number of cycles or time.

The presence and duration of the Phase 2 interval are intricately linked to both the load amplitude and the resulting heat generation within the material. In some cases, Phase 2 may be entirely absent, observed when the load amplitude reaches such high levels that the transition from Phase 1 bypasses Phase 2 entirely, progressing directly to Phase 3, and culminating in the eventual fracture of the specimen.

These three characteristic phases of temperature evolution are commonly observed in metallic structures, mainly analysed for steel alloys [13; 14; 3; 4; 15; 16], due to the presence of a theoretical fatigue limit. Similarly, for aluminium alloys [17], fatigue limit transitions are present.

An alternative response to cyclic loading is observed in certain materials, such as pure copper [18]. While these materials also exhibit a temperature evolution separated into three phases, Phase 2 deviates from the constant temperature response seen in other materials. Instead, the temperature in Phase 2 fails to stabilise, and a non-zero temperature increase rate becomes apparent (see Fig. 2.). This unique behaviour adds an additional layer of complexity to the cyclic loading responses, showcasing the diverse nature of the material responses under different loading conditions.

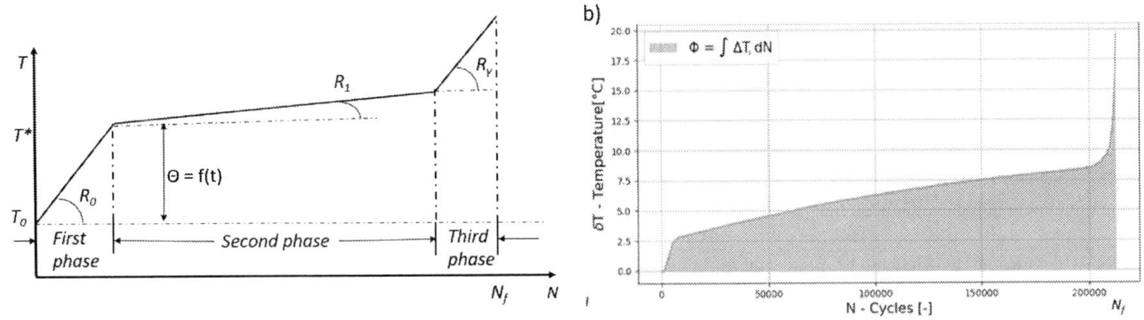

Fig. 2 Schematic representation of three phases of the temperature response during constant-amplitude cyclic loading until N_f at failure, but without any stabilised region in Phase 2.

This paper introduces a modified approach to the prediction of fatigue life, based on Fargione's method [3], with a specific emphasis on the self-heating effect. In particular, our study addresses cases where the temperature evolution lacks Phase 2, as observed in previous research [19]. Five configurations of conventionally manufactured steel were investigated, each with varying surface finishes, leading to various fatigue life performances.

2 Experimental Campaign

This experimental campaign focused on investigating the impact of various machining parameters on the fatigue life performance of the test series A25-29. These series were produced from conventionally manufactured steel 42CrMo4+QT, originating from a single heat source and fabricated from a 35mm diameter bar. Part of a larger campaign conducted throughout Europe, this study builds on previous research exploring the effect of size [20; 21] and self-heating [19]. The ultimate tensile strength and yield stress of the material were determined to be 1097 MPa and 1001.5MPa, respectively. Variations in surface roughness across series A25-A29, influenced by machining parameters used for the finishing cut detailed in the Tab. 1, resulted in different fatigue strengths and thermal responses during fatigue testing. In particular, differences in the final finish were characterized by the roughness achieved and residual stresses on the surface (see Fig. 4). All series were manufactured using the same lathe, with the fatigue specimen geometry depicted in Fig. 3 for all series A25-A29 made from 42CrMo4+QT steel.

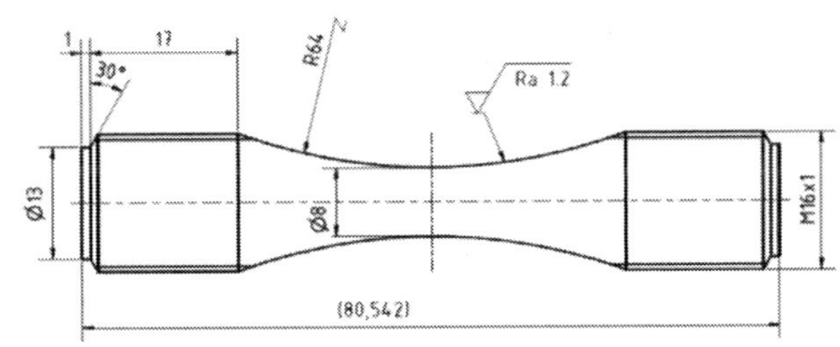

Fig. 3 Geometry of fatigue specimens manufactured from 42CrMo4+QT steel, within the project [22; 23; 19].

All fatigue experiments were conducted on an Amsler resonant (see Fig. 5) pulsator at CTU in Prague with a load ratio of R = -1 and testing frequency of around 150Hz. This difference in fatigue performance of each series is attributed to the variation of production parameters, leading to high residual stresses on the surface of the specimens (see Table 1, and the legend of the graph in Fig. 4), particularly noticeable in tension for all series as so as the roughness of final finish. In particular, A28 shows the poorest fatigue performance, despite a similar S-N curve trend ranging from 10^4 to 10^5 for both A28 and A26. Series A27 and A25 exhibit nearly identical fatigue curve regressions using the K&V regression method [24]. Furthermore, the K&V regression of A29 and A26 demonstrates a nearly identical transition to the fatigue limit domain. The lower fatigue performance of A28 could also be attributed to the highest measured roughness (see Tab. 1). The fatigue limit was established as an asymptote to the K&V regression [24].

In this paper, the focus is set on self-heating and fatigue life estimation derived from the self-heating effect. Consequently, the detailed description of the fatigue strength corresponding to various machining parameters is described in another publication by Papuga et al. [25].

Tab. 1: Different turning parameters used for machining the fatigue samples for series A25-A29. All series were turned without additional grinding or polishing.

		A25	A26	A27	A28	A29
Cutting speed at D=8mm	V_c [m/min]	75	75.4	100	62.8	75.4
Feed rate	F [mm/rev]	0.05	0.05	0.1	0.1	0.1
Process fluid (Decocut 1040 during roughing)	Y/N	Y	Y	Y	Y	Y
Cooling fluid during last cuts	Y/N	Y	N	Y	N	N
Cutter nose radius	R [mm]	0.4	0.4	0.4	0.4	0.4
Last cut height	[mm]	0.05	0.1	0.1	0.05	0.1
Second-last cut height	[mm]	0.1	0.1	0.1	0.1	0.1
Third-last cut height	[mm]	0.15	0.1	0.1	0.15	0.1
Ra	[μm]	1.26	0.67	1.2	1.46	0.87
Rz	[μm]	6.4	3	5.4	7.1	3.3

Self-heating experiments were performed using thermal camera Flir A315 (with the *NETD* of 50mK) with surface of the samples covered with high emissivity painting from LabIR (HERP-LT-MWIR-BK-11) to improve the precision of the measurement. Near the loaded specimen, a reference non-loaded specimen was positioned. Serving as a stable temperature reference, this unloaded sample played a crucial role in our measurements. When the thermal differentials between the loaded and unloaded samples were compared, the heat generation within the loaded specimen could be assessed more accurately.

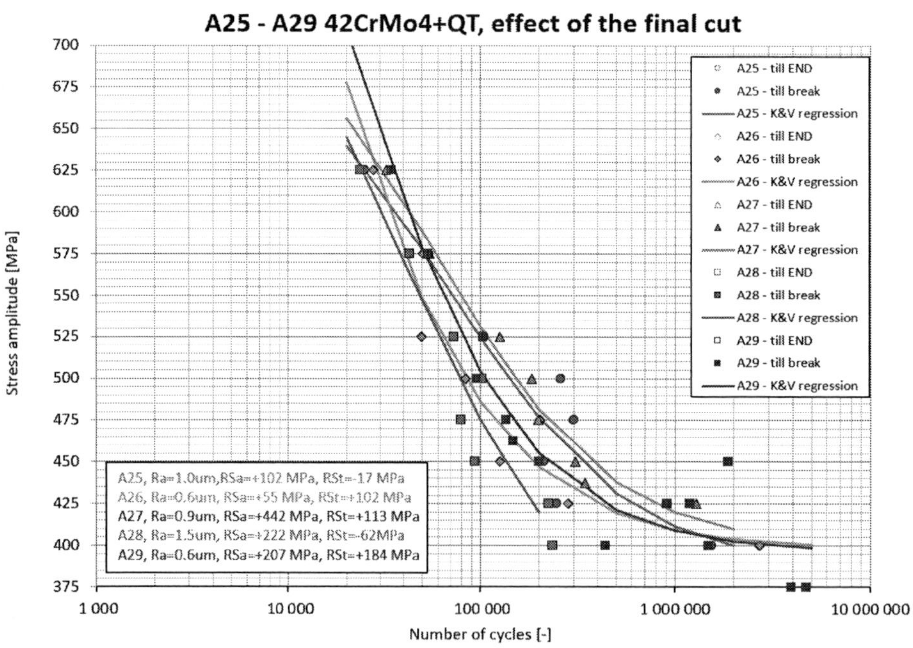

Fig. 4 Fatigue life curves with K&V regression for A25-A29. The *RSa* parameter refers to the residual stress in the axial direction, and *RSt* is the residual stress in the tangential direction.

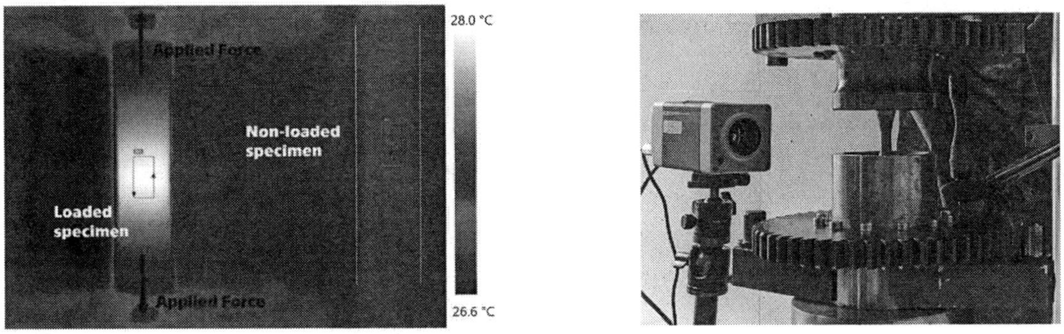

Fig. 5 Thermal image (FLIR A315 320×240p, with frame rate 3.75 Hz) of the reference specimens and loaded (left). Amsler resonant pulsator with a FLUKE RSE600 thermal imaging camera. The loaded and reference specimens are installed in the machine (right).

3 The Fargione method

The method proposed by Fargione [3], also utilized by Lipski [26], offers an intriguing analysis of the relationship between applied stress and the number of cycles to failure. Central to this method is the concept of "limiting energy“ derived from the second law of thermodynamics. This principle states that a body can generate only a certain amount of entropy before structural failure occurs, essentially representing the maximum disorderliness achievable. Fargione defines this energy as the integration of the temperature evolution curve over the interval of cycles from the start of the measurement to the point of failure. He suggests that Phases 1 and 3 are negligible, leading to the definition of an area as a function of the number of cycles to failure and the stabilised temperature increase, Eq. (1). This concept is sometimes referred to as "finite energy" or "limiting energy," representing a parameter associated with the finite entropy of the system; see Fig. 6.

$$\Phi = \int_0^{N_f} T(N)\, dN = const. \tag{1}$$

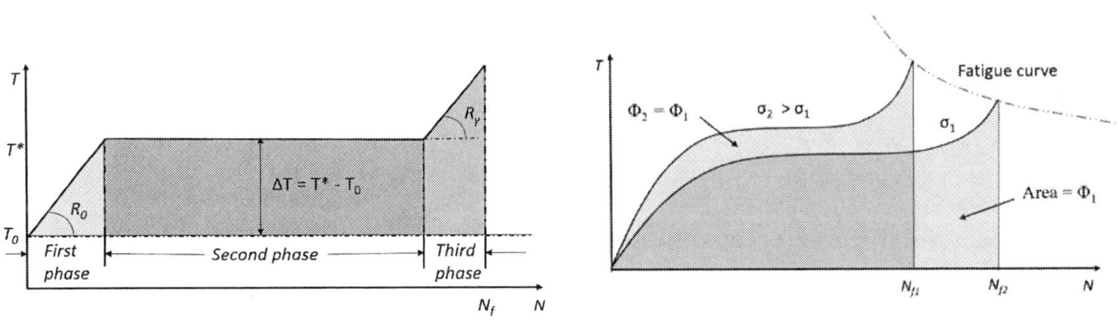

Fig. 6 The figure on the left represents the expected temperature evolution for the use of the Fargione method. The right scheme illustrates Fargione’s method with estimating the number of cycles to failure for each amplitude of load with the use of the concept of the limiting energy.

Fargione's approach [3] expands into thermographic analysis, providing deeper insight into material behaviour under cyclic loads compared to Risitano's method [14]. The goal is not only to determine the fatigue limit, but also to characterise the entire S-N curve. Component failure under dynamic loads is attributed to microplasticity, leading to increased yielding. In this context, a limiting energy is introduced to describe the dissipated energy throughout the failure process. The heat energy which is directly proportional to the limiting energy Φ can be interpreted as the integral of the $\Theta(N)$ function. Fargione postulates that this limiting energy remains constant, serving as a characteristic for the material or component.

Rearranging the formula, the number of cycles until fracture (N_f) can be determined from the observed stabilised temperature increase Θ. Since Θ is characteristic of the applied stress amplitude according to Risitano [14], N_f can be correlated with the stress amplitude σ_a through this relationship. The characteristic stabilised temperatures are obtained from the step test (see Fig. 7), allowing the numbers of cycles N_f to be associated with every corresponding stress amplitude σ_a. This method requires a step test for multiple stress levels and, optionally, a pure $\Theta(N)$ curve from the onset to the failure (single-level test). The stress amplitude should be well above the fatigue limit to determine the limiting energy of the single-level test, ensuring that the test is ended by a failure of the specimen. While theoretically achievable with a single sample, at least three samples are recommended for calculating an average value of limiting energy, as well as for the S-H step-test.

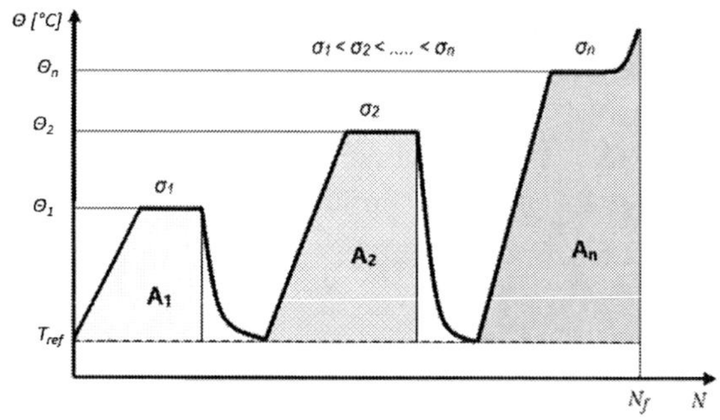

Fig. 7 Scheme of the Self-Heating (S-H) step test with the step amplitudes subsequently increasing in each step and depicted limiting energy areas A_i within each loading block.

As Eq. (1) relies on the entire temperature evolution, it needs to be simplified. Our goal is to identify a measurable parameter, specifically the stabilised temperature in Phase 2, that can be consistently monitored across sequential loadings on a single sample during the self-heating step-test. Fargione recommends using the 1st and 2nd phases, as stated in Eq. (2)

$$\Phi \approx \frac{1}{2} \cdot \Theta \cdot N_{1-2} + \Theta \cdot (N_f - N_{1-2}) = const. \qquad (2)$$

where the stabilised temperature Θ represents the temperature increase in Phase 2, denoted by Θ. The value of $N_{1\text{-}2}$ corresponds to the number of cycles at the transition from Phase 1 to Phase 2. Phase 3 is ignored, as it cannot be accounted for in the SH step test with a variable amplitude. To simplify Eq. (2), Phase 1 is neglected, as it plays a minor role in the limiting energy parameter, as stated in [3] in certain cases. Fargione's limiting energy parameter is then simplified to:

$$\Phi \approx \Theta \cdot N_f = const. \qquad (3)$$

3.1 Modification of the Fargione method for special cases of 42CrMo4+QT

In this section, an improved version of Fargione's method was developed to address the non-standard thermal response observed in special cases. The same batch of 42CrMo4+QT was examined in other articles [19; 23; 27]. However, the primary focus here is on the S-H effect, with reference [19] highlighting a critical issue: the absence of the stabilised temperature increase in Phase 2, as illustrated in Fig. 2. For this paper, the fatigue limit (FL) is defined as the value obtained as an asymptote to the K&V regression [24], mainly because most FL estimation methods rely on temperature stabilisation in the second phase. Despite testing various parameters of the temperature evolution, no results directly related to FL were obtained [19].

Determining the limiting energy typically involves cyclically loading a single specimen at a specific stress amplitude until failure occurs. If Fargione's assumption, as expressed in Eq.(1), holds true, this limiting energy value can be utilised to estimate the number of cycles to failure at other stress levels, provided that the stabilised temperature response to the corresponding load amplitudes is acquired. It is crucial to acknowledge that relying solely on a single limiting energy value can be misleading, as it disregards the statistical nature of fatigue damage, which follows a distribution rather than a precise value.

An intriguing application of Fargione's theory, as outlined in Eq. (1), involves the experimental determination of the S-N curve using only two test specimens. One specimen is utilised to obtain the limiting energy value, while the second specimen undergoes testing with progressively increasing stress amplitudes, as depicted in Fig. 6 (right). In the specific cases addressed here, up to nine samples from constant-amplitude tests intended to determine the S-N curve were utilized to calculate the limiting energy. This approach allows for a deeper understanding of whether the limiting energy remains constant or varies with the stress amplitude.

In the case of the 42CrMo4 + QT tests described earlier; it was observed that there is no section of the stabilised temperature increase [19]. Another assumption had to be made because the Fargione method is based on reaching the stabilised Phase 2. This acknowledges the dynamic nature of the testing process and aims to incorporate any initial changes and subsequent stabilisation in Phase 2, ensuring a more accurate estimation of fatigue behaviour.

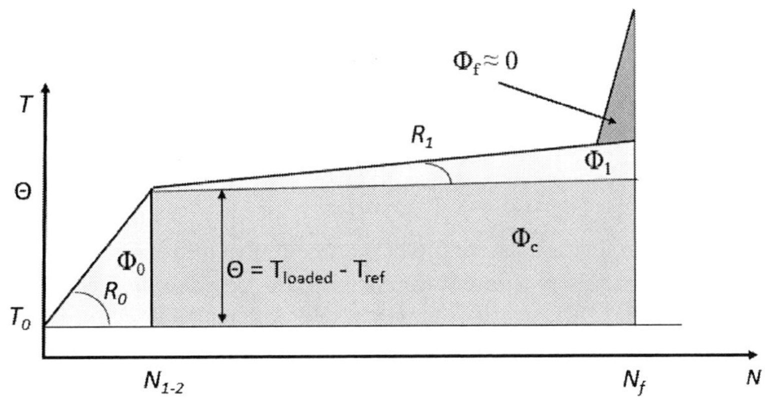

Fig. 8 Four segments of the limiting energy. The sum of all partial limiting energies represents the limiting energy of the whole temperature evolution.

Fig. 8 presents a schematic representation that illustrates the separation of the limiting energy into four distinct areas of focus. Each of these areas contributes to the overall limiting energy of the entire temperature evolution. By considering the cumulative sum of these four areas, the limiting energy can be obtained. To implement the Self-Heating step-test with variable load amplitude, which comprises only Phase 1 and Phase 2 of temperature evolution during cyclic loading, the segmentation assumes a linear behaviour in the second phase. This approach allows for the estimation of the limiting energy for the entire temperature evolution based on a small segment of the total lifespan, thereby predicting the number of cycles to failure.

The main change involves setting the limiting energy from the measured parameters of the Θ -N record obtained from a step test, using the $N_{1\text{-}2}$ which is the number of cycles at the end of the first phase, the secondary temperature increase rate in Phase 2 R_1 and the temperature Θ which is related to the original Fargione method described in Fig. 2. In this paper, limiting energy is assumed to be a function of the load amplitude as observed by Lipski [26] or in [19; 5]. To demonstrate the relationship between stress amplitude and failure limits of the tested specimens at a single level, see Fig. 9. It is evident that as stress amplitude increases, the limiting energy decreases, highlighting the variable nature of this parameter in this specific scenario.

The area most related to the original Fargione's assumption Eq. (3) is Φ_c (referred to as c as its main parameter is stabilized temperature Θ) that can be described as:

$$\Phi_c \approx \Theta(N_f - N_{1-2}), \tag{4}$$

where Φ_c represents a part of the limiting energy as depicted in Fig. 8. $N_{1\text{-}2}$ represents the length of Phase 1 in cycles, and it is set as an intersection of the record segments with rates of R_0 and R_1. Another segment is the area in Phase 1 described as Φ_0:

$$\Phi_0 \approx \frac{1}{2}\Theta N_{1-2}, \tag{5}$$

The main reason for the modification of Fargione's original approach is the R_1 rate in Phase 2, which creates a new partial limiting energy segment Φ_1:

$$\Phi_1 \approx \frac{1}{2}\tan R_1 \left(N_f - N_{1-2}\right)^2 . \tag{6}$$

Eq. (6) can be further simplified so that the tangent of an angle, which is nearly zero, is equal to the angle itself. Since Phase 3 does not occur in the self-heating step test with a variable load amplitude, the partial limiting energy Φ_f (f as an area close to failure of the structure) is neglected in this model, but it is supposed to be nearly equal to zero. Using the Eq. (4) to (6), the limiting energy of the entire test is then described as follows:

$$\Phi \approx \Theta\left(N_f - N_{1-2}\right) + \frac{1}{2}\Theta \cdot N_{1-2} + \frac{1}{2}\tan R_1 \left(N_f - N_{1-2}\right)^2 . \tag{7}$$

It is also assumed that the limiting energy is not constant across various stress amplitudes, but that it is constant for the same stress amplitude in the same testing condition and for the same material. It is assumed that the function of the limiting energy to the stress amplitude is a power-law function:

$$\Phi \approx f(\sigma_a)\,, \Phi \approx a_1\, \sigma_a^{\,a_2}. \tag{8}$$

Coefficients a_1 and a_2 in Eq. (8) are determined experimentally from single-level tests to failure of the specimen. For the series used in this experiment campaign, A25-A29, the coefficients are established using regression, as shown in the Tab. 2. It is worth noting that the coefficients of determination are relatively low, primarily due to the scatter of data, as depicted in Fig. 9.

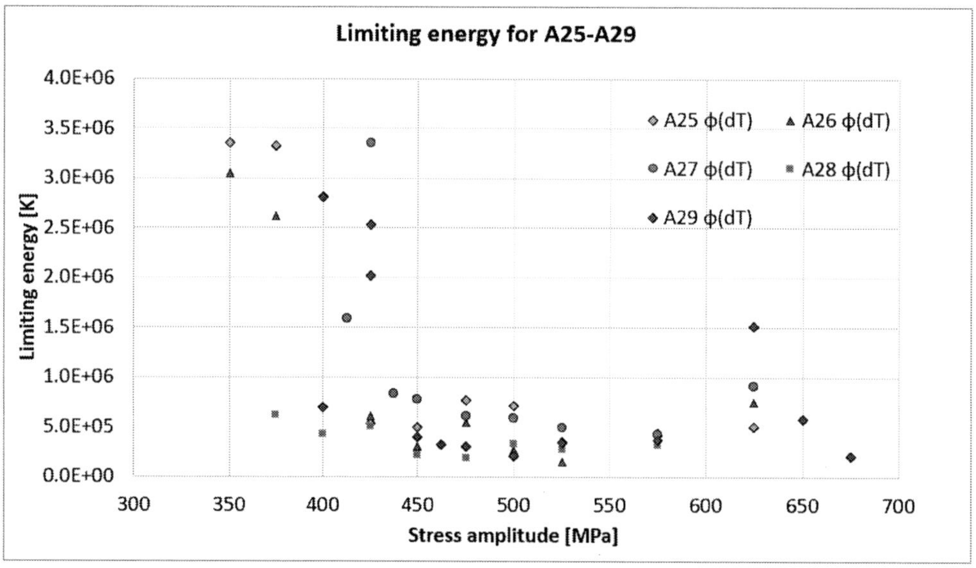

Fig. 9 Relation between stress amplitude and limiting energy calculated according to single level test for series A25-A29.

Tab. 2: Coefficients for describing the limiting energy via the power-law function Eq. (8).

	a_1	a_2	R^2
A25	2.93E+16	-3.95	0.81
A26	9.90E+16	-4.17	0.48
A27	9.02E+06	0.00	0.31
A28	8.60E+09	-1.66	0.52
A29	8.92E+16	-4.09	0.53

The number of cycles is then set as follows using Eq. (7) and (8):

$$N_f \approx \frac{1}{2}\left(\left(2N_{1-2} - \frac{\Theta}{2R_1}\right) + \sqrt{\left(-2N_{1-2} + \frac{\Theta}{2R_1}\right)^2 - 4\cdot\left(N_{1-2}^2 - \frac{\Theta N_{1-2}}{4R_1} - \frac{\left(\frac{\sigma_a}{a_1}\right)^{1/a_2}}{2R_1}\right)}\right) \quad (9)$$

3.2 Application of the modified Fargione method

First, the limiting energy is calculated, and the coefficients of the power law Eq. (8) are established as outlined in the Tab. 2. Subsequently, using Eq. (9), the number of cycles is determined for each stress amplitude in the S-H step test. This calculation requires input parameters such as N_{1-2}, R_1, Θ (obtained from the self-heating step test) and σ_a from the load cell of the Amsler resonant pulsator.

The self-heating step test was performed to determine N_{1-2}, R_1, and Θ, with a focus on stress amplitudes well below the fatigue limit for each configuration. The test block comprised 15,000 cycles per stress amplitude, with the stress amplitudes progressively increasing until failure. The primary objective was to estimate the S-N curve, when the conventional Fargione approach using stabilized temperature was not applicable. In the absence of a lower limit of usability of Fargione's method, it is assumed that the fatigue limit is the lower boundary [28]. In this study, the fatigue limit is determined from the data obtained from the experimentally derived S-N curve (refer to Fig. 4). Each fatigue curve estimation utilised a single sample, except for A29, where three samples could be used thanks to the bigger number of specimens available.

In Fig. 10, the results of the modified Fargione method show a strong correlation with the K&V regression of series A26, A27, and A29, with A29 and A26 exhibiting particularly accurate predictions for the estimated number of cycles to failure. However, without establishing a fatigue limit (FL) as an asymptote to K&V regression [24], this estimate could extend to the lowest value tested, underscoring the need for limitations, which were not identified in this study. Series A28 shows significantly higher fatigue life predictions compared to the experimentally established slope of the S-N curve, suggesting potential inaccuracies in component production. Series A25 shows good correspondence in the upper part of the S-N curve up to $2\cdot10^5$ cycles, with trends nearly identical to A27 but with closer alignment to the experimentally measured S-N curve. However, beyond this point, the trend takes a steeper slope than the measured S-N curve, potentially influenced by an outlier specimen, as observed in the A28 series. The coefficient of determination (R^2) for the A25-A29 series estimates ranges from 0.30 to 0.78, with the lowest R^2 observed for series A28, as discussed earlier.

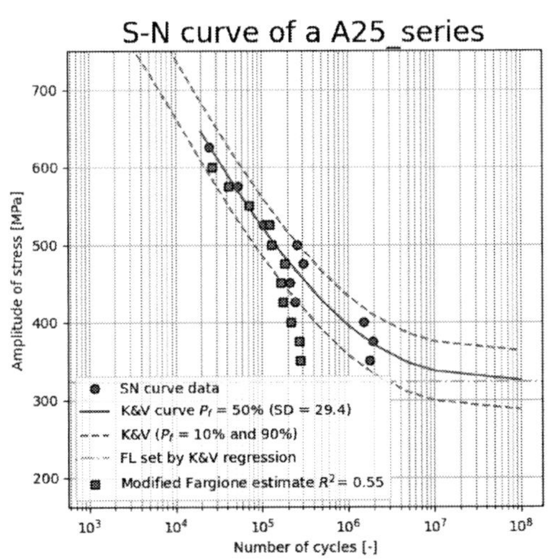

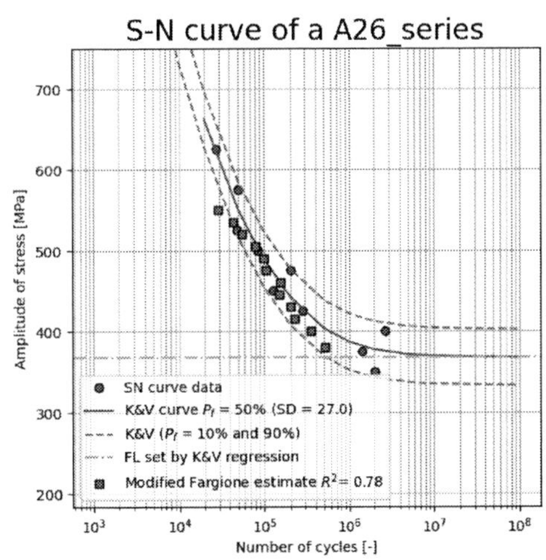

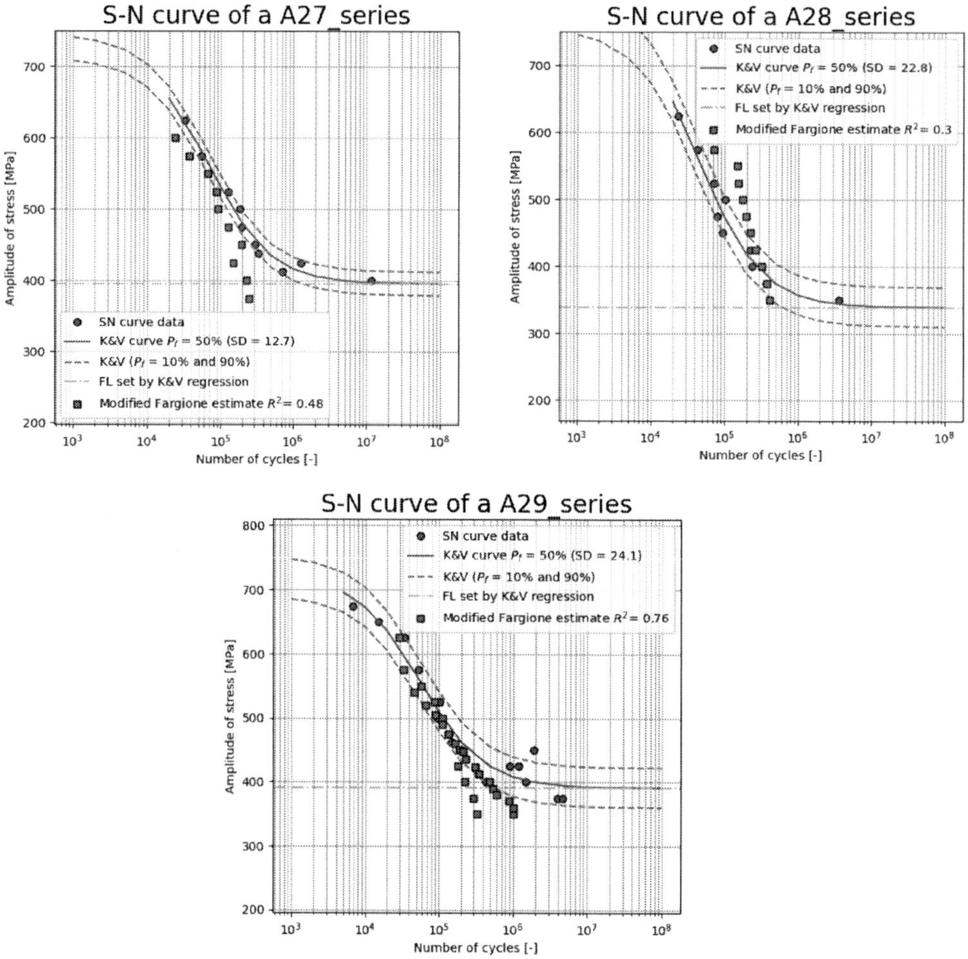

Fig. 10 Modified Fargione method for series A25-A29 with marked estimates and coefficients of determination in each figure.

Undoubtedly, the main drawback of the modified Fargione method lies in the variable limiting energy, as indicated in Tab. 2. The weak correlation of the limiting energy with different stress amplitudes reveals a poor fit to the dataset, reflected in the low coefficient of determination values. Another challenge in predicting the number of cycles to failure stems from the observed shifts in trends, particularly noticeable in A25 and A27 in the modified Fargione estimation. Although this change could potentially result from damage incurred during previous testing levels, such occurrences should ideally be minimised. On the contrary, the single specimen from the A28 series might represent an outlier on the positive side of the S-N curve, leading to potentially misleading estimates. Addressing these issues is paramount for refining this method for presentation purposes.

4 Conclusion

This study presents a comprehensive investigation of fatigue life assessment, focussing on the unique case of the 42CrMo4+QT material and its unconventional temperature response during cyclic loading. Using the self-heating effect and modifying Fargione's method, a promising approach to predicting fatigue life is proposed. The experimental campaign on five datasets highlights the influence of machining parameters on fatigue life and residual stresses, providing insights into material behaviour, however, the main focus of this paper is not on the influence of machining parameters of 42CrMo4+QT steel.

The results of the thermographic observation indicate that this material lacks the typical stabilisation of the temperature of Phase 2 commonly observed. The proposed modification of the Fargione method addresses the abnormal Phase 2 without temperature stabilization, showing promising results in predicting fatigue life.

However, challenges such as variable limiting energy and shifts in trends pose significant considerations for the accuracy and reliability of the method. The application of the modified Fargione approach works partially for the series, where only one sample was dedicated for S-H step-test A25-A28, leading to more conservative results for A25 and A27. Series A28 has its estimate much higher than the S-N curve, which could potentially be caused by using a specimen, that would be an outlier on the right side of the fatigue curve. For future research, employing at least three specimens for the S-H step-test from which the average thermal and fatigue behaviour is obtained, as demonstrated in the A29 series, appears to be the most promising approach.

Acknowledgement

The authors acknowledge the support by the Grant Agency of the CTU in Prague (SGS23/156/OHK2/3T/12 project) and by the Czech Ministry of Education, Youth and Sports (LUABA22071 project).

References

[1] Fatemi A, Yang L. Cumulative fatigue damage and life prediction theories: a survey of the state of the art for homogeneous materials. International Journal of Fatigue 1998;vol. 20:9-34. https://doi.org/10.1016/S0142-1123(97)00081-9.

[2] Standard Practice for Statistical Analysis of Linear or Linearized Stress-Life (S-N)and Strain-Life (e-N) Fatigue Data 1981.

[3] Fargione G. Rapid determination of the fatigue curve by the thermographic method. International Journal of Fatigue 2001;vol. 24:11-19. https://doi.org/10.1016/S0142-1123(01)00107-4.

[4] Meneghetti G. Analysis of the fatigue strength of a stainless steel based on the energy dissipation. International Journal of Fatigue 2007;vol. 29:81-94. https://doi.org/10.1016/j.ijfatigue.2006.02.043.

[5] Matušů M, Dimke K, Šimota J, Papuga J, Rosenthal J, Mára V, Beránek L. Energy-based method for analyzing fatigue properties of additively manufactured AlSi10Mg. Journal of Mechanical Science and Technology 2023;37:7. https://doi.org/10.1007/s12206-022-2110-6.

[6] Seitl S, Klusák J, Pelayo F, Canteli A. Thermographic Determination Methodology: Application on Fatigue Limit of AL 2024 for R=-1. Key Engineering Materials 2013;577-578:477-480. https://doi.org/10.4028/www.scientific.net/KEM.577-578.477.

[7] Klesnil M, Lukáš P. Fatigue of metallic materials. 3. upr. vyd. Praha: Academia; 1992.

[8] Matušů M, Džuberová L, Papuga J, Rosenthal J, Šimota J, Beránek L. Fatigue analysis and heat treatment comparison of additively manufactured specimens from AlSi10Mg alloy. International Journal of Fatigue 2024;vol. 185:18. https://doi.org/10.1016/j.ijfatigue.2024.108357.

[9] Guo Q, Zaïri F, Yang W. Evaluation of intrinsic dissipation based on self-heating effect in high-cycle metal fatigue. International Journal of Fatigue 2020;vol. 139. https://doi.org/10.1016/j.ijfatigue.2020.105653.

[10] Amiri M, Khonsari M. Rapid determination of fatigue failure based on temperature evolution: Fully reversed bending load. International Journal of Fatigue 2010;vol. 32:382-389. https://doi.org/10.1016/j.ijfatigue.2009.07.015.

[11] Prochazka R, Dzugan J, Konopik P. Fatigue limit evaluation of structure materials based on thermographic analysis. Procedia Structural Integrity 2017;vol. 7:315-320. https://doi.org/10.1016/j.prostr.2017.11.094.

[12] Douellou C, Balandraud X, Duc E, Verquin B, Lefebvre F, Sar F. Fast fatigue characterization by infrared thermography for additive manufacturing. Procedia Structural Integrity 2019;vol. 19:90-100. https://doi.org/10.1016/j.prostr.2019.12.011.

[13] Luong M. Fatigue limit evaluation of metals using an infrared thermographic technique. Mechanics of Materials 1998;vol. 28:155-163. https://doi.org/10.1016/S0167-6636(97)00047-1.

[14] La Rosa G. Thermographic methodology for rapid determination of the fatigue limit of materials and mechanical components. International Journal of Fatigue 2000;vol. 22:65-73. https://doi.org/10.1016/S0142-1123(99)00088-2.

[15] Meneghetti G, Ricotta M. The use of the specific heat loss to analyse the low- and high-cycle fatigue behaviour of plain and notched specimens made of a stainless steel. Engineering Fracture Mechanics 2012;vol. 81:2-16. https://doi.org/10.1016/j.engfracmech.2011.06.010.

[16] Cura F, Curti G, Sesana R. A new iteration method for the thermographic determination of fatigue limit in steels. International Journal of Fatigue 2005;vol. 27:453-459. https://doi.org/10.1016/j.ijfatigue.2003.12.009.

[17] Ezanno A, Doudard C, Calloch S, Millot T, Heuzé J. Fast characterization of high-cycle fatigue properties of a cast copper–aluminum alloy by self-heating measurements under cyclic loadings. Procedia Engineering 2010;vol. 2:967-976. https://doi.org/10.1016/j.proeng.2010.03.105.

[18] Wang X, Crupi V, Jiang C, Feng E, Guglielmino E, Wang C. Energy-based approach for fatigue life prediction of pure copper. International Journal of Fatigue 2017;vol. 104:243-250. https://doi.org/10.1016/j.ijfatigue.2017.07.025.

[19] Matušů M, Papuga J, Mžourek M. Fatigue strength estimation of 42CrMo4 QT from the temperature evolution during cyclic loading. Procedia Structural Integrity 2022;vol. 42:102-109. https://doi.org/10.1016/j.prostr.2022.12.012.

[20] Papuga J, Mžourek M, Matušů M, Mára V, Čapek J. Investigation of the size effect on 42CrMo4 QT steel in the high-cycle fatigue domain part I: Experimental campaign. International Journal of Fatigue 2023;vol. 175. https://doi.org/10.1016/j.ijfatigue.2023.107743.

[21] Mžourek M, Papuga J, Nesládek M, Matušů M, Čapek J, Mára V. Fatigue Damage Analysis On 42CrMo4 QT Via Critical Volume Approach. Procedia Structural Integrity 2022;vol. 42:457-464. https://doi.org/10.1016/j.prostr.2022.12.058.

[22] Mžourek M, Papuga J, Nesládek M, Matušů M, Čapek J, Mára V. Fatigue Damage Analysis On 42CrMo4 QT Via Critical Volume Approach. Procedia Structural Integrity 2022;vol. 42:457-464. https://doi.org/10.1016/j.prostr.2022.12.058.

[23] Papuga J, Mžourek M, Matušů M, Mára V, Čapek J. Investigation of the size effect on 42CrMo4 QT steel in the high-cycle fatigue domain part I: Experimental campaign. International Journal of Fatigue 2023;vol. 175. https://doi.org/10.1016/j.ijfatigue.2023.107743.

[24] Kohout J, Věchet S. A new function for fatigue curves characterization and its multiple merits. International Journal of Fatigue 2001;vol. 23:175-183. https://doi.org/10.1016/S0142-1123(00)00082-7.

[25] Papuga J, Vrbata T, Trojan K, Trefil A, Zabala A, Mára V. The Effect of Specimen Machining Processes on Fatigue Life Of 42CrMo4+Qt Steel. 9th EIS International Conference on Durability & Fatigue – Fatigue 2024 (In - print), Cambridge: Jesus Coledge, UK; 2024, p. 13.

[26] Lipski A. Rapid Determination of the S - N Curve for Steel by means of the Thermographic Method. Advances in Materials Science and Engineering 2016;vol. 2016:1-8. https://doi.org/10.1155/2016/4134021.

[27] Mžourek M, Papuga J, Nesládek M, Matušů M, Čapek J, Mára V. Fatigue Damage Analysis On 42CrMo4 QT Via Critical Volume Approach. Procedia Structural Integrity 2022;vol. 42:457-464. https://doi.org/10.1016/j.prostr.2022.12.058.

[28] Matušů M, Šimota J, Papuga J, Rosenthal J, Reis L, Costa P, Bumba F. Estimation of the S-N curve from limiting energy for additively manufactured specimens from AlSi10Mg. Fatigue & Fracture of Engineering Materials & Structures [Under Review] n.d.:32.

On a Possibility of High-Resolution Neutron Scattering Observations of TiAl Substructure

P. Mikula [1,a], A. Michalcová [2], V. Ryukhtin [1]

[1] *Nuclear Physics Institute ASCR, 250 68 Řež, Czech Republic*
[2] *Department of Metals and Corrosion Engineering, University of Chemistry and Technology, 166 28 Prague 6, Czech Republic*
[a] *mikula@ujf.cas.cz*

Abstract: In this contribution, we present the feasibility of using two unconventional neutron scattering methods, namely, high-resolution neutron diffraction and small-angle neutron scattering, for substructure studies of TiAl alloys. The first one permits the finer studies of the diffraction line profiles and the second one the size distribution of pores on a rather large scale. The feasibility experiments were carried out on the three TiAl samples of the $\phi = 9$ mm diameter which were made from the 3D printed material of gamma titanium aluminides (Ti48Al2Cr2Nb) loaded on the bulk Ti substrate.

Keywords: neutron diffraction; small-angle scattering; TiAl alloy; porosity; plastic deformation.

1 Introduction

Thanks to their excellent properties (e.g. low density, superior strength, creep, and oxidation resistance, the γ-TiAl alloys are considered as one of the most promising materials fulfilling the demands for next-generation high-temperature materials in the aerospace, automotive and energy industries etc. [1, 2]. The material density of γ-TiAl alloys is approximately half of the conventional Ni-based superalloys, their use can significantly reduce the weight of aircraft and automobile engines, and therefore, also to decrease the emission of greenhouse gases, as well as reduce fuel consumption. It is clear that much attention is paid to the investigation of the influence of the technological preparation of the TiAl alloys on their properties strongly dependent, naturally, on their structure. In our contribution, we present the results of the first studies of the high-resolution neutron powder diffraction profiles and porosity by small-angle neutron scattering (SANS) on three samples of TiAl alloy in relation to their technological preparation. However, the problem could be that Ti and Al have an opposite sign of the neutron scattering amplitude making a relatively small structure factor for diffraction on the crystalline grains resulting in a low detector signal.

2 Experimental Details and Results

The TiAl samples of the cylindrical form, all of a diameter of ϕ=9 mm, were prepared. The manufacture of specimens was carried out by using a printing device InssTek MX-LAB producing the spot size, the height of the layer and the hatch distance. Laser power, speed of the printing and dwell time were 130 W, 849 mm/min and 3.5 sec, respectively. Parameters related to the samples were: Both S1 (a preheat of the platform on 500 °C) and S2 (without a preheat) - the 800 mm spot size, the 150 mm height of the layer using the cladding head SDM 800 and the 500 mm hatch distance and S3 (no preheat) - the 400 mm spot size, the 250 mm height of the layer using the cladding head SDM 400 and the 300 mm hatch distance.

2.1 Neutron Diffraction

The neutron diffraction experiment was carried out by a high-resolution three-axis setting as schematically shown in Fig. 1 [3-5]. Following the sketch (for small widths of the samples), a maximum resolution can be achieved for minimal dispersion of the whole system which can be achieved by the optimization of the curvature of the bent perfect crystal monochromator and analyzer [3]. The diffraction profiles are obtained by rocking the BPC analyzer and registering the detector signal for individual angular steps.

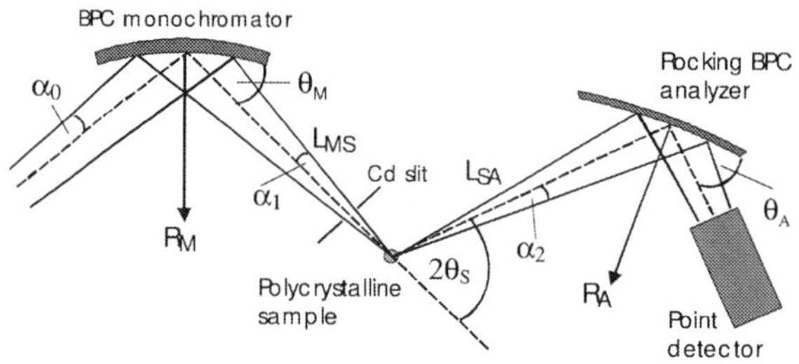

Fig. 1: Three-axis diffractometer setting employing BPC monochromator and analyzer as used in the experiment (R_M, R_A - radii of curvature, θ_M, θ_A - Bragg angles and L_{MS} and L_{SA} - distances).

2.2 Small-Angle Neutron Scattering (SANS)

Figure 2 shows the scheme of the high-resolution SANS diffractometer with a PSD detector [6, 7]. It is designed for the measurements of neutron small-angle scattering in the high Q-resolution mode. In contrast to the conventional double-crystal arrangements, the fully asymmetric diffraction geometry on the elastically bent Si analyzer is employed in order to transfer the angular distribution of the scattered neutrons to the spatial distribution. Then, it enables to analyze the whole scattering curve by a one-dimensional position sensitive detector. The instrument is mainly suited to the investigation of structural or compositional inhomogeneities in materials in the size range of 0.05 µm – 2 µm, mainly porous materials and large precipitates in alloys. The remote control of the curvatures of the monochromator and analyzer crystals makes it possible to tune the instrument resolution easily in the Q range from $2x10^{-3}$ to $2x10^{-1}$ nm^{-1}, according to the expected size of investigated inhomogeneities.

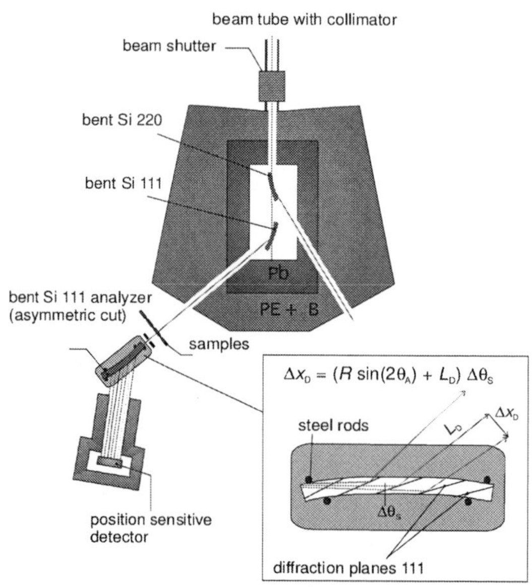

Fig. 2: Schematic diagram of the high-resolution double-bent-crystal SANS diffractometer [6, 7]. The inserted part shows the transformation of the scattering angular deviation on the position coordinates.

2.3 Experimental Results

2.3.1 Diffraction Analysis

First, a powder diffraction spectrum was taken on one of the samples, in order to see the basic structure of the material (see Fig. 3). For this we used medium-resolution powder diffractometer MEREDIT [8] also installed at the Řež's reactor. The analysis of the spectrum confirmed the existence of the single γ phase of the TiAl alloy. Then, we analyzed the second peak with the high-resolution setting. Figure 4 shows the analyser rocking curves demonstrating how the technology preparation of the samples could reflect their *FWHM*s.

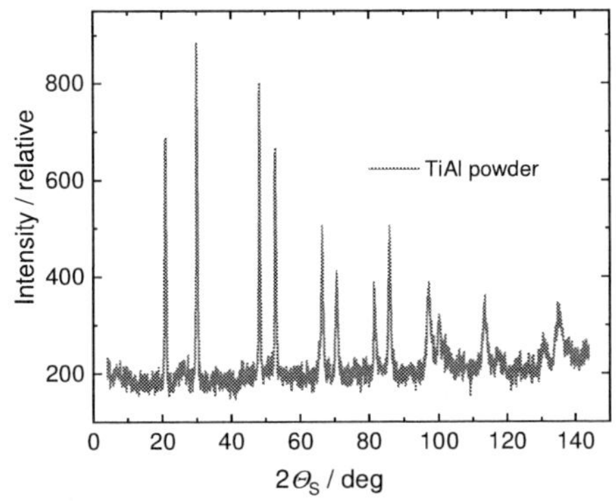

Fig. 3: Powder diffraction spectrum taken on the TiAl powder sample.

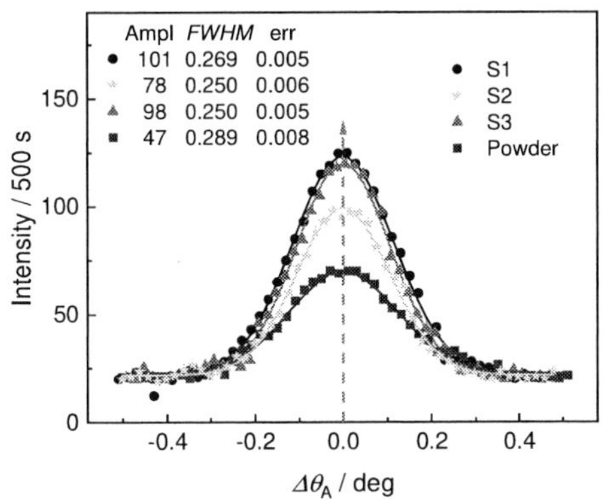

Fig. 4: Analyzer rocking curves related to the individual TiAl samples of the diameter of ϕ = 9 mm: (a) – sample S1, (b) – sample S2, (c) – sample S3 and (d) – powder sample.

2.3.2 SANS Analysis

Figure 5 shows the size-distribution functions $D(R)$ of the pores obtained by the analysis of the SANS data.

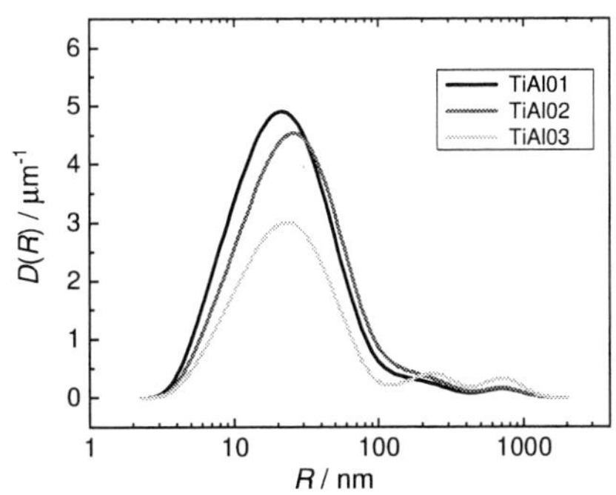

Fig. 5: Size-distribution functions of pores for the individual samples as obtained by the SANS method and analysis.

3 Conclusion

Three chosen TiAl samples were prepared by the printing device InssTek MX-LAB from the powder material and investigated by the high-resolution three-axis diffractometer when looking for the *FWHMs* of diffraction profiles and then, by the neutron small-angle scattering instrument looking for the size distribution of the contained pores as a result of an influence of the technology preparation. Inspection of Fig. 4 reveals that *FWHM*s related to the samples S2 and S3 are equal within the experimental error which points out that the used sample preparation has nearly the same influence on their plastic deformation. However, the sample S2 was poorly compacted and crumblier. On the other hand, it has a rather strong influence on the size distribution of pores (see Fig. 5). In this way it should be mentioned that the former instrument is also suitable for studies of polycrystalline alloys containing several phases having close lattice spacings as was demonstrated earlier on Inconel 718 alloy [9] while the second instrument is generally suited for investigation of structural or compositional inhomogeneities in materials in the size range from 0.05 μm to 2 μm, namely, porous materials and large precipitates in alloys. Finally, it can be stated that both presented high-resolution neutron-scattering methods can offer additional useful support to complement the structural information achieved by using the other conventional characterization methodologies (e.g. with electron microscopy techniques) and can be interesting for potential external users who are dealing with the structure of TiAl alloys and would like to apply for a beam time by means of the CANAM project at some of the instruments [10].

Acknowledgements

Measurements were carried out at the CANAM instrument of NPI CAS Řež installed at the

CICRR infrastructure, which is financially supported by the Ministry of Education and Culture - project LM2023041. The authors acknowledge support from ESS participation of the Czech Republic – OP (CZ.02.1.01/0.0/0.0/16_013/0001794) and from the project ESS Scandinavia-CZ II (LM2018111), respectively. Furthermore, they acknowledge support from the Czech Academy of Sciences in the frame of the program "Strategie AV21, No. 23". The authors thank B. Michalcová from NPI CAS for significant help with measurements and data elaborations.

References

[1] C. Shen, K.D. Liss, M. Reid, Z. Pan, Y. Ma, X. Li, H. Li, In-situ neutron diffraction characterization on the phase evolution of γ-TiAl alloy during the wire-arc additive manufacturing process, Journal of Alloys and Compounds 778 (2019) 280–287, https://doi.org/10.1016/j.jallcom.2018.11.150.

[2] G. Németh, Investigation of titanium alloys using neutron diffraction, PhD Thesis, Prague, Faculty of Mathematics and Physics, Charles University Prague, 2021, http://hdl.handle.net/20.500.11956/128397.

[3] P. Mikula, J. Šaroun, V. Ryukhtin, High-resolution residual strain/stress measurements on three-axis neutron diffractometer, Advanced Materials Research 1166 (2021) 33–40. https://doi.org/10.4028/www.scientific.net/AMR.1166.33, https://www.ujf.cas.cz/en/departments/department-of-neutron-physics/instruments/lvr15/hk8b/.

[4] P. Mikula, J. Šaroun, V. Ryukhtin, J. Stammers, An alternative neutron diffractometer performance for strain/stress measurements, Powder diffraction 35(3) (2020) 185–189, https://doi.org/10.1017/S0885715620000329.

[5] P. Mikula, V. Ryukhtin, High-resolution three-axis neutron diffractometer for microstructure investigations of polycrystalline materials, Procedia Structural Integrity 43 (2023) 119–123, https://doi.org/10.1016/j.prostr.2022.12.245.

[6] J. Šaroun, P. Lukáš, P. Mikula, P. Strunz, Focusing doubly bent crystal diffractometer in combination with PSD for SANS experiments, Journal de Physique, Coll. C7, 3 (1993) 439–442.

[7] P. Lukáš, P. Mikula, J. Šaroun, P. Strunz, Double bent crystal diffractometer for SANS experiments, Nuclear Instruments and Methods in Physics Research A 338 (1994) 111–115, https://www.ujf.cas.cz/en/departments/department-of-neutron-physics/instruments/lvr15/hk8a/.

[8] Retrieved from: https://www.ujf.cas.cz/en/departments/department-of-neutron-physics/instruments/lvr15/ meredit/instrument-description.html.

[9] P. Mikula, V. Ryukhtin, On a possibility of high-resolution neutron diffraction observation of Inconel 718 substructure phases, in proc.: 61st conference on experimental stress analysis EAN (2023), book of extended abstracts, eds. M. Hagara, E. Huňady, Technical University of Košice, Košice, Slovakia.

[10] Retrieved from: https://www.ujf.cas.cz/cs/vyzkum-a-vyvoj/open-access/.

Shrinkage of Rammed Unfired Earth in Long Term Observation

P. Padevět [1,a], B. Mužíková [1]

[1] *Czech Technical University in Prague, Faculty of Civil Engineering, Department of Mechanics, Thákurova 7, Prague 6, 166 29, Czech Republic*
[a] *pavel.padevet@fsv.cvut.cz*

Abstract: The paper is devoted to the creep and shrinkage of compacted unfired earth. The creep of illitic-kaolinitic clay has been monitored for a long time. The initial measurement was focused on the initial course of creep. The long-term measurement took place for almost 4 years. The compacted clay was prepared using a water coefficient of 0.295, when the amount of illitic-caolinitic clay used in the mixture was 30%.

Keywords: rammed unfired earth; illitic-caolinitic clay; shrinkage; creeping of material; water ratio.

1 Introduction

Rammed unburned earth has been a traditional building material since time immemorial. In recent centuries, however, modern materials such as concrete, steel and fired ceramic products have been used. Rammed earth has receded into the background of modern materials [1]. However, there are still applications where unburned earth is useful in building structures. A weakness at present is the little knowledge of the properties of rammed unburned earth needed for the design of load-bearing structures.

For construction purposes, rammed unfired earth is a material for which it is advisable to focus not only on strength properties, but also on rheological characteristics. An important factor for monitoring the rheology of unfired clay is the need for mixing water during the creation of the mass. Creep and shrinkage are rheological properties of clay that have a significant effect on the long-term behavior of the structure.

The goal of creep measurement is to determine the behavior of the material over time. The rheological properties of compacted clay and their measurement are dependent on the knowledge of the material properties, especially the compressive strength. Knowing the material properties allows us to correctly set the amount of load for measuring creep. Fresh compacted clay increases its compressive strength by gradual drying, i.e. the release of loosely bound water, which is necessary for high-quality processing during the preparation of the mixture.

2 Tested Material

The pressed clay was prepared from sand, illitic kaolinitic clay and water. The mass ratio of water to illitic kaolinitic clay was 0.295. The ratio of sand and clay was 70:30. The samples for creep and shrinkage testing were prepared together with the samples intended for material tests.

The ratio of water to total solids of the mixture was 0.4. The ratio is expressed by the coefficient w - water coefficient. The amount of solids is the sum of the weights of clay and sand used. Sand plays the role of filler in the mixture, clay has the function of binder. The binding function of clay consists of covering the grains of sand with a layer of clay. It binds the water in the mixture and creates a suitable binder structure of the mixture. If the water content in the mixture is lower, the necessary binding structure of the binder is not formed. On the contrary, if the amount of water in the mixture is higher than the amount needed to create the binding capacity of the clay, the water remains outside the binding space. This water disappears during drying and after the free water, an imperfectly dense - more porous structure of compacted clay is formed. The tested mixture belonged to the area with a higher ratio of water and clay.

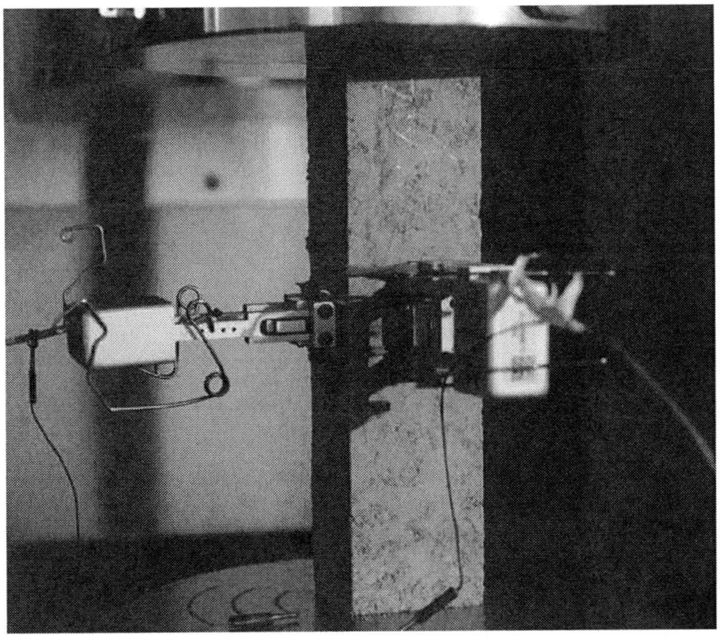

Fig. 1: Illitic kaolinitic clay prism specimen in compression test with sand clay ratio 70 : 30.

The material properties were investigated on prisms 40 × 40 × 160 mm Fig.1., which were produced simultaneously with the bodies for creep. The creep specimens had a hexagon-shaped base with a distance of 20 mm from opposite edges. The height of the specimens was 70 mm due to the installation in the lever measuring mechanism, as is seen in Fig. 2. The measurement was started on 6 specimens, where 3 samples were intended for creep measurement and 3 samples for shrinkage measurement.

Fig. 2: One of the specimens of illitic kaolinitic clay after creep test.

The production of both types of samples took place by making a mixture of clay, sand and water. The mixture was placed in layers in steel molds and compacted [2]. After filling the mold and compacting the last layer, the specimens were removed from the molds and left to dry naturally in laboratory conditions. The bodies intended to measure creep and shrinkage were immediately placed in the lever mechanisms and the measurement was started.

3 Instrumentation of Experiment

The creep specimens had a hexagon-shaped base with a distance of 20 mm from opposite edges. The height of the specimens was 70 mm. All tests took place in lever mechanisms. A lever mechanism for measuring creep applied a load to the specimen of 53 N. Shrinkage was measured in a lever device that was used only to clamp the strain sensors. These specimens were not subjected to force, as in the case of creep measurements. The size of creep and shrinkage was recorded using optoelectronic probes SM 12. 3 sensors were placed around each specimen in the axial direction of loading of the samples, see Fig. 2.

The specimens were immediately inserted into the test devices and started with the measurement. The data has been collected in a locally placed PC file. Creep measurement was carried out on 3 specimens. Shrinkage was measured on two specimens. One specimen that was intended for creep measurement was damaged during instrumentation. A total of 5 specimens were used for measurement.

During the measurement period, the temperature at the place where the samples were tested was recorded. During the first 4 months, measurements were recorded regularly. In the following period, the measurement was recorded sporadically, while the location of the electrical supply of the probes was maintained throughout the measurement period. However, the actual data storage was very sporadic as time progressed. Data recording during regular measurements was carried out every 5 minutes.

The continuous measurement period was 1557 days. However, the data recording was sporadic, as can be seen in Fig. 3, where the evolution of the temperature during the measurements in the laboratory is shown in Fig. 3.

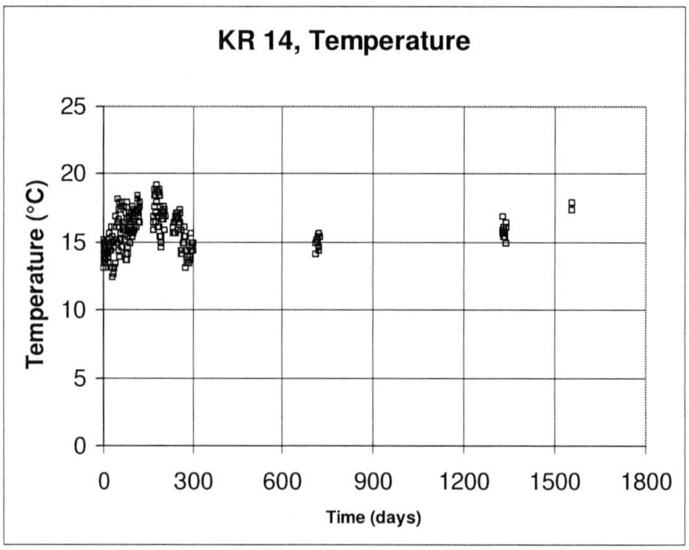

Fig. 3: Evolution of temperature during the testing.

4 Results

Figures 3 to 6 show measured data from a long-term measurement. The method of connecting the measured values into a curve was not used for the display, but only the measured values in the graphs would be left. The data has been reduced so that the creep and shrinkage values are shown in the graphs in the number of one record per day. In the initial phase, the measured values are more frequent.

The temperature varied between 14 and 17 °C during the measurement. The oscillation that is visible in the first year of measurement can be explained by the increased temperature between autumn and spring. The beginning of the measurements was carried out in the summer months. Heating during the winter months in the laboratory was important for changing the temperature. If we were to use all the recorded values, we

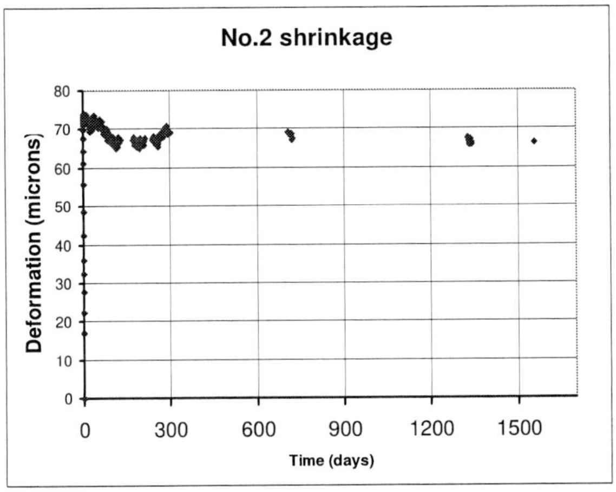

a) Shrinkage curve of rammed earth prepared from illitic kaolinitic clay, water ratio 0.4

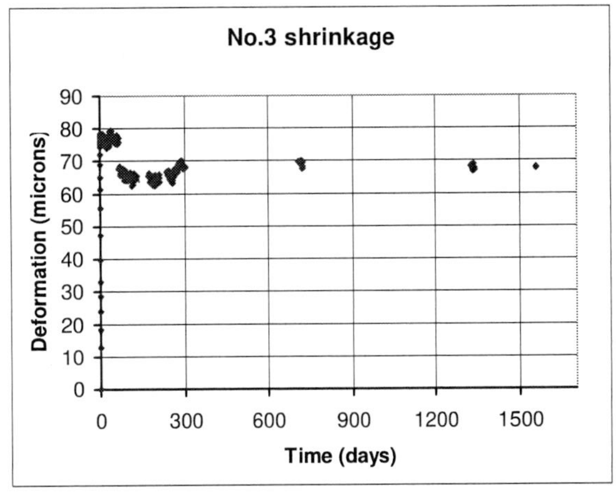

b) Shrinkage curve of rammed earth prepared from illitic kaolinitic clay, water ratio 0.4

Fig. 4: Curves of shrinkage for rammed unfired earth with ratio of clay and sand 30 : 70 and water ratio 0.4.

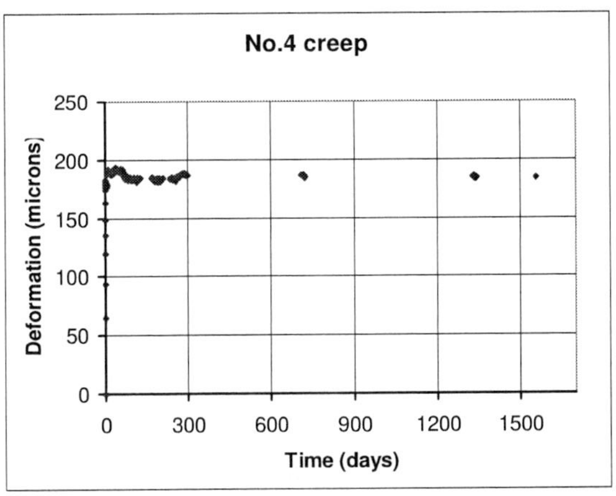

a) Creep curve of rammed earth prepared from illitic kaolinitic clay, water ratio 0.4

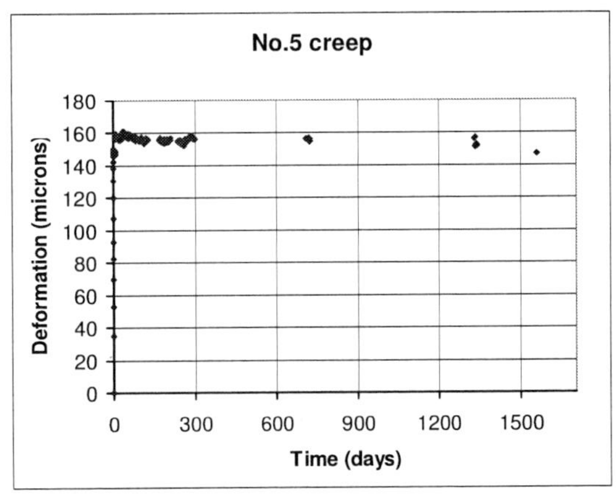

b) Creepe curve of rammed earth prepared from illitic kaolinitic clay, water ratio 0.4

Fig. 5: Curves of creep for rammed unfired earth with ratio of clay and sand 30 : 70 and water ratio of 0.4.

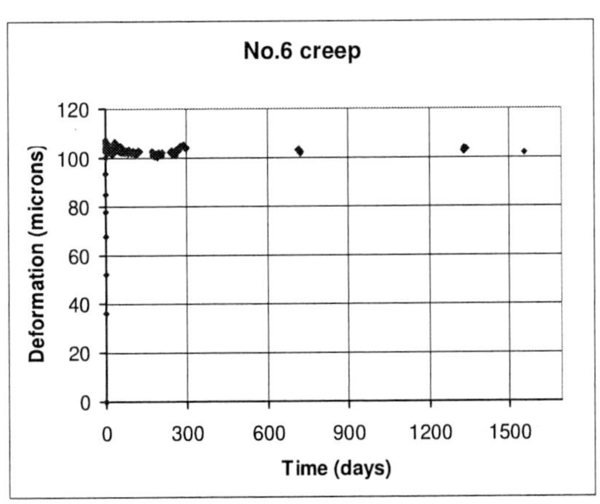

Fig. 6: Creep curve of rammed earth prepared from illitic kaolinitic clay, water ratio of 0.4.

would find that the temperature fluctuation with a difference of 3°C is more likely due to the daily change of temperatures and the ability of the air conditioner to regulate the temperature change during the day. The temperature change is comparable to, for example, the temperature change recorded during a similar measurement [3].

Shrinkage that occurred during the measurement is most noticeable at the initial time of data recording. At the beginning, the material dries out, which is accompanied by a change in the thickness of the specimens. In both cases of the specimens shown in Figure 4, the shrinkage value was around 70 microns at a length of 70 mm. This means that the relative strain has a value of $1 \cdot 10^{-3}$.

After drying, i.e. evaporation of free bound water and constant temperature in the laboratory, the shrinkage remained at the achieved value of 70 microns.

In the case of the creep measurement, the evolution of the deformation was similar to the deformation from shrinkage. The graphs in Figures 5 and 6 show an increase in deformation at the beginning of the measurement and the subsequent steady development of the measurement throughout the measurement period.

The largest deformation values were achieved at 180, 158 and 103 microns. The highest values were recorded around the 40th day of measurement. Subsequently, the size of the deformation did not increase. This is in contrast to the shrinkage measurements where the highest values were reached on the second day from the start.

5 Conclusion

The result of the measurement is a comparison of the creep and shrinkage of the crushed adobe material, which contained a significant amount of mixing water. The very plastic material resulted in damage to one specimen during instrumentation in the measuring device, as well as considerable deformation of the material at the beginning of the measurement. The result of the experiment can be summarized in the following points:

- The size of the water coefficient has a significant effect on the amount of shrinkage and creep of the material.
- Shrinkage was 50% of the average creep value of unfired clay (70 microns / 147 microns).
- After reaching strength, i.e. evaporation of loosely bound water, there is no significant change in the amount of creep and shrinkage even during long-term observation.
- Illitic kaolinitic clay in the mixture has a stable mixture behavior. The amount of creep was significantly affected by the large water factor. For less saturated mixtures, the amount of creep is up to 10 × smaller. After reaching the strength, the deformation of the material hardly changes.

Acknowledgement

The paper was supported by the project CTU No. SGS 24/037/OHK1/1T/11.

References

[1] G. Minke, Building With Earth [online], [published August 18, 2015], ISBN 978-3-0346-0822-0, available from: http://archive.org/details/Gernot Minke-Building With Earth 2006.

[2] B. Mužíková, T. Plaček Otcovská, P. Padevět, Modulus of Elasticity of Unfired Rammed Earth, Acta Polytechnica CTU Proceedings 15 (2018) 63–68, https://doi.org/10.14311/APP.2018.15.0063.

[3] P. Padevět, P. Bittnar, B. Mužíková, Rheological Properties of Rammed Clay from Illitic Clay in Relation to the Amount of Illite, 4th International Scientific Conference Structural and Physical Aspects of Construction Engineering (SPACE 2019), Matec Web of Conference 310 (2020) 6 pages, https://doi.org/10.1051/matecconf/202031000057.

Analysis of the strain rate dependence on the mechanical properties of polypropylene

P. Palička [1,a], R. Huňady [1], M. Hančinová [1], E. Dubňanská [1]

[1] *Technical University of Košice; Letná 1/9, Košice, Slovakia; peter.palicka@tuke.sk, robert.hunady@tuke.sk, martina.hancinova@tuke.sk, erika.dubnanska@tuke.sk*
[a] *peter.palicka@tuke.sk*

Abstract: Since the end of the 20th century, the use of plastics has undoubtedly been expanding, not only in technical applications, but also in food industry, medicine, and many other sectors. The mechanical behaviour of polymeric materials is significantly influenced by temperature, strain-rate and their processing. This paper investigates the effect of strain rate on the mechanical properties of polypropylene filled with 12% talc. Bending, DMA and uniaxial tensile tests at three strain rates were carried out as part of the material testing. The obtained stress-strain data were adjusted to align Young's modulus with the storage modulus by modifying the curve slopes. This methodology can be applied for evaluating mechanical properties to input into material models for FEM simulations.

Keywords: polypropylene; viscoelasticity; strain rate; mechanical properties; mechanical testing

1 Introduction

The same parameters are used to characterise the mechanical properties of plastics as for metals, i.e. Young's modulus, Yield stress, Ultimate strength, Poisson's ratio, but their mechanical behaviour differs considerably to metals. In comparison to other materials, the mechanical response of plastics varies to a much greater extent depending on the strain rate, temperature, and the external environment [1-3]. Polymers at high strain rate or low temperature behave like glassy substances, while at low strain rate or high temperature polymers behave like rubbery materials. In the first case, the mechanical properties are higher, but the polymer becomes brittle, while in the second case the mechanical properties are lower and the polymer becomes ductile [4].

The investigated material was polypropylene filled with 12% talc (PP_TD12). Polypropylene is a semi-crystalline thermoplastic produced by the polymerisation of propylene. Polypropylene has high strength and stiffness. There is a wide range of different applications for polypropylene, from automotive parts to packaging foils [5][6].

Dynamic mechanical analysis (DMA) is a method for determining mechanical properties of viscoelastic materials. Applying an oscillating (sinusoidal) stress to a material, its response can be measured as function of frequency. From phase lag between applied stress and measured strain is determined tendency of material to flow (viscosity), while the sample recovery reflects material's stiffness (modulus). Since modulus equals to stress/strain ratio, the complex modulus can be calculated. Then the storage modulus and loss modulus can be derived as components of complex modulus using measured phase lag between two sine waves. The storage modulus is the elastic component, and it is directly related to the stiffness of the sample. Loss modulus is the viscous component, which is dependent on the material's ability to dissipate mechanical energy through molecular motion [7].

2 Experimental investigation

Polypropylene PP_TD12 was tested in two perpendicular directions, 0° and 90° according to the cutting orientation from the base plastic plate made by injection. It was necessary to verify whether the material can be considered homogeneous and isotropic due to the talc filler. Fig. 1 shows tensile specimens and Fig. 2 shows the specimens for DMA and bending tests.

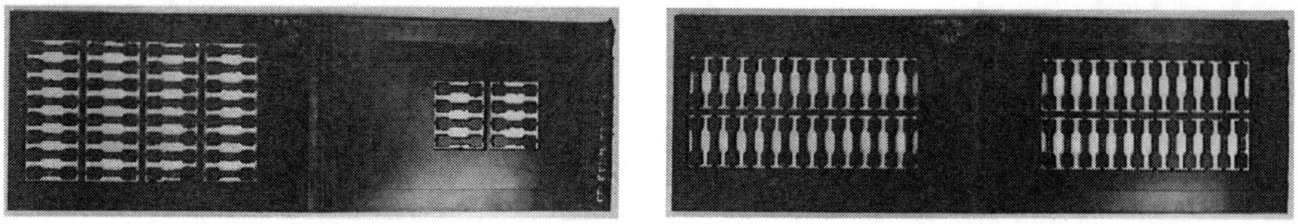

a) b)

Fig. 1 Tensile test specimens: a) 0°; b) 90°.

a) b)

Fig. 2 DMA and bending test specimens: a) 0°; b) 90°.

Uniaxial tensile tests were performed on the universal machine LabTest 5.050ST at strain rates of 10^{-2}, 10^{0} s^{-1}. Impact tensile test on drop tester was used as technique for measurement of high strain rates and specimens was loaded with strain rate 10^{2} s^{-1}. Seven specimens was measured for each strain rate and results was averaged to obtain representative curves. Tensile tests were carried out to obtain material characteristics (stress-strain curves) up to the failure. Three-point bending tests were performed on the same machine as tensile tests to determine the flexural modulus at a strain rate of 10^{-2} s^{-1}. This test was performed to verify the results of the DMA tests. DMA analysis was performed on a DMA Q800 device in a three-point bending test at four frequencies of 0.04, 0.4, 4.0, and 40 Hz to find the frequency dependence of the storage modulus. Bending tests, including DMA, was realized on five specimens.

The test specimen for the tensile tests was in the shape of a conventional "dogbone" specimen with a thickness of 2.5 mm and the specimen for the flexural and DMA tests was in the shape of a rectangle with the same thickness. The specimens were small in size to avoid some undesirable effects due to the high loading rates. The tensile, flexural and DMA tests are shown in Fig. 3.

a) b) c)

Fig. 3: Material tests: a) uniaxial tensile, b) three-point bending, c) DMA.

Fig. 4 shows the average stress-strain curves evaluated from tensile tests for both specimen orientations. From the plots can be seen that curves correspond well in the elastic region but differ in the plastic region at failure.

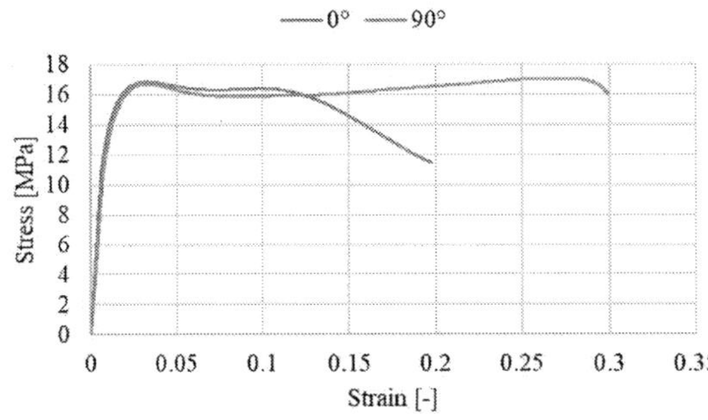

a)

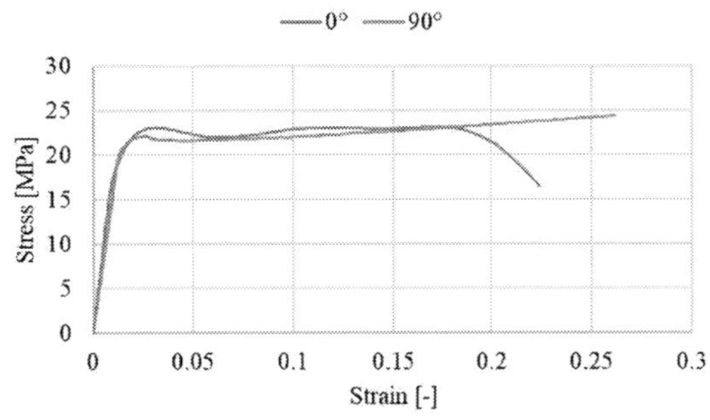

b)

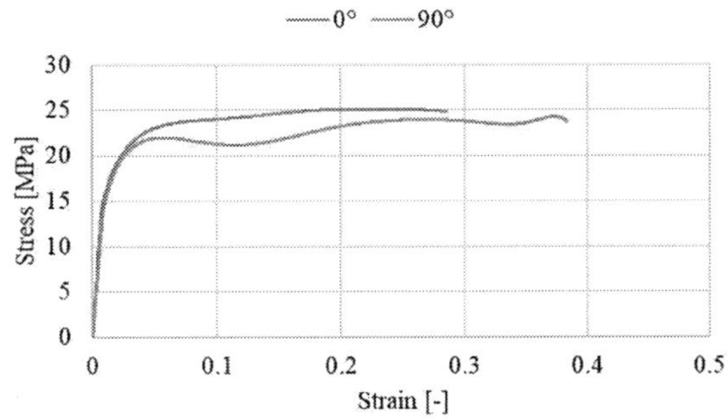

c)

Fig. 4 Average stress-strain curves measured at strain rate: a) 10^{-2} s^{-1}; b) 10^{0} s^{-1}; c) 10^{2} s^{-1}.

In Fig. 4c, can be seen that the strain value at failure is too high, since the samples were measured at the highest strain rate. During impact tensile test on drop tester, only the force as a function of time was measured. It means that the strain values were calculated only as an approximation from the measured time data and the strain rate. For these tests, the curves show significant oscillation. This oscillation is caused by the propagation of the shock wave from the ram impact on the lower grip. The measured curves had to be smoothed and approximated. They were approximated in Matlab software using polynomial functions. For demonstration, evaluated engineering stress-strain curve and the approximated curve from the impact tensile test are shown in Fig. 5.

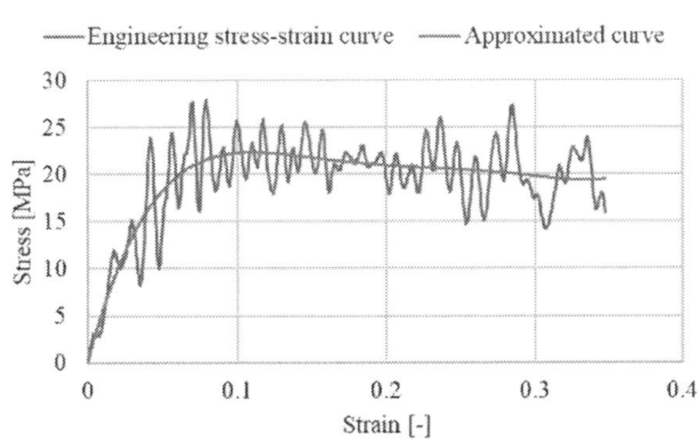

Fig. 5 Engineering and approximated stress-strain curve.

Since the orientation of the talc filler in injection molded components can be arbitrary with respect to the loading direction, the stress-strain curves for 0° and 90° were averaged and the material was further treated as homogenized isotropic. For the same reason, the modulus values obtained from the DMA tests were also averaged. Tab. 1 shows the measured and average value of flexural modulus. The resultant mean flexural modulus is 1599 MPa. The resultant flexural modulus value is in good agreement with the result from the DMA analysis, where the measured storage modulus for a frequency of 0.4 Hz is 1621 MPa. This frequency corresponds approximately to the loading rate of the three-point bending test. The average values of storage modulus and loss modulus measured at the other frequencies are shown in Tab. 2.

Tab. 1 Measured and average value of flexural modulus obtained by three-point bending test.

Orientation	0°					90°					Mean
Specimen	PP_01	PP_02	PP_03	PP_4	PP_5	PP_1	PP_2	PP_3	PP_4	PP_5	
Flexural modulus [MPa]	1730	1715	1755	1836	1757	1447	1420	1477	1449	1400	1599

Tab. 2 Average values from DMA test.

Log. Frequency [Hz]	Frequency [Hz]	Storage modulus E' [MPa]	Loss modulus E'' [MPa]
-1.4	0.04	1490	138
-0.4	0.4	1621	113
0.6	4	1756	96
1.6	40	1849	141

The frequency dependence of storage modulus can be seen in Fig. 6, where it is obvious that as the frequency increases, the value of storage modulus also increases.

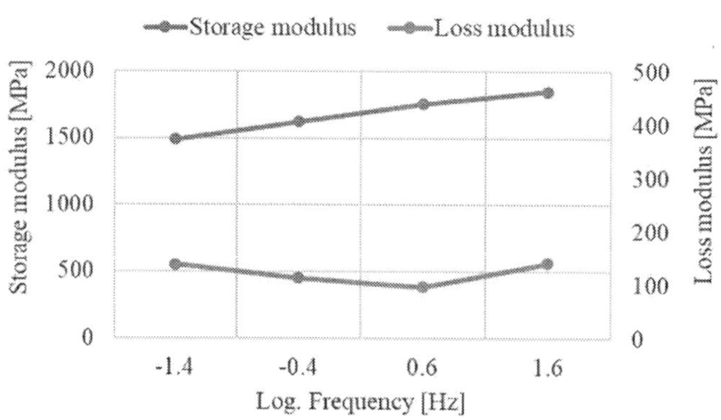

Fig. 6 Frequency dependence of storage modulus.

As it was mentioned, from impact tensile test the strain values were calculated only as an approximation from the measured time data. It means, that the elasticity modulus (slope of the stress-strain curve) is not correct and it needs to be corrected. Storage modulus determined from DMA tests describes elastic region of these curves. By using the following methodology, it is possible to directly adjust the tensile curve to the storage modulus. The method is the following:
The total strain ε_t of the engineering stress-strain curve can be expressed as

$$\varepsilon_t = \varepsilon_e + \varepsilon_{pl}\,, \tag{1}$$

where ε_e is elastic strain of specimen and ε_{pl} is plastic strain. From the stress-strain curve, only the plastic strain needs to be expressed according to the equation

$$\varepsilon_{pl} = \varepsilon_t - \frac{\sigma}{E}, \tag{2}$$

where σ is stress value and E is Young's modulus from tensile test. Using equation (2), all elastic strain values are set to zero, leaving only the plastic strain values. With equation

$$\varepsilon_t = \varepsilon_{pl} + \frac{\sigma}{E'}, \tag{3}$$

where E' is storage modulus, the elastic deformation corresponding to the storage modulus is added to plastic deformation from previous equation. These steps result in a modified tensile curve for the storage modulus. The initial averaged engineering stress-strain curve (blue) with $E = 556$ MPa and the stress-strain curve (red) with storage modulus $E' = 1849$ MPa modified according the eq. (3) are shown in Fig. 7.

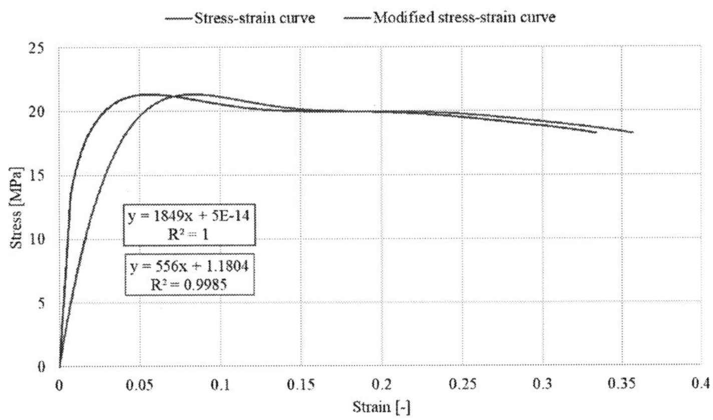

Fig. 7 Tensile and modified stress-strain curve.

In material models used in FEM simulations, it is necessary to transform engineering stress-strain curves to true curves using the volume conservation law

$$\varepsilon_{true} = \ln(1 + \varepsilon_e), \qquad \sigma_{true} = \sigma_e(1 + \varepsilon_e). \tag{4}$$

This law is valid only up to the ultimate strength. The resultant true stress-true strain curve is shown in Fig. 8.

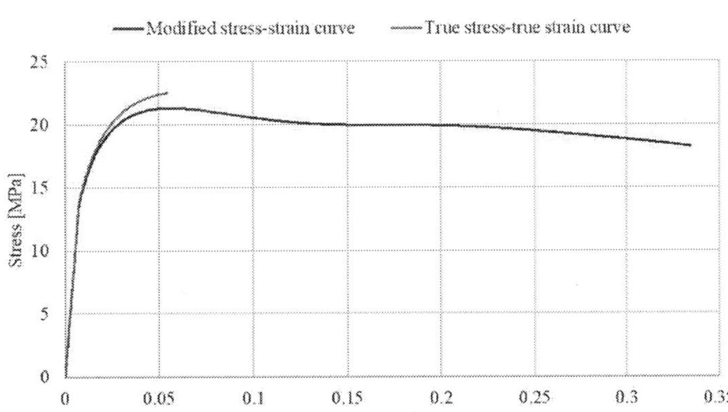

Fig. 8 True stress-true strain curve in comparison with engineering stress-strain curve.

In Fig. 9, the dependence of strain rate on the stress-strain curves can be seen. Due to the increase of strain rate, the material becomes stronger and stiffer. These curves describing the region up to the ultimate strength can be applied in material models for FEM simulations.

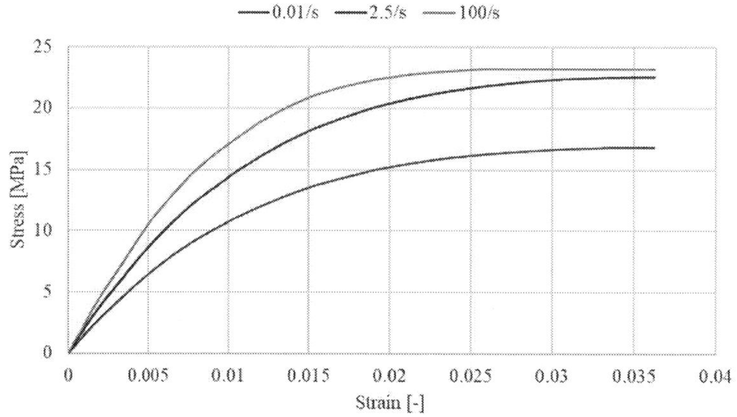

Fig. 9 Resultant true stress-true strain curves for different strain rates..

3 Conclusion

The aim of this paper was to investigate the effect of strain rate on mechanical properties of polypropylene filled with 12% talc. To accomplish the aims of the paper, three types of mechanical tests were performed. The tensile test was carried out to obtain the stress-strain characteristic. The three-point bending test was used as a validation tool for the DMA tests, where the frequency dependence of the storage modulus was measured. The impact tensile tests at a strain rate of 10^2 s^{-1} measured on the drop test machine had to be approximated for further use. After obtaining the stress-strain characteristics for all strain rates, it was further necessary to adjust the Young´s modulus to the storage modulus, i.e. to adjust the slope of the curves. This whole process of evaluating the mechanical properties of the polymer and further curve modifying can be used as a methodology for evaluating mechanical properties for input into material models for FEM simulations.

Acknowledgement

This work was supported by the Science Grant Agency of the Ministry of Education, Science, Research and Sport of the Slovak Republic and the Slovak Academy of Sciences under the project VEGA 1/0516/22.

References

[1] B. Jiang; L. Cao; F. Zhu. Dynamic tensile behaviour of polypropylene with temperature effect, Composites Part B: Engineering, Volume 152, 2018, pp. 300-304.

[2] M. Zrida et al., High-speed tensile tests on a polypropylene material, Polymer Testing, Volume 29, Issue 6, 2010, pp. 685-692.

[3] P.A. Du Bois et al., Material behaviour of polymers under impact loading, International Journal of Impact Engineering, Volume 32, Issue 5, 2006, pages 725-740.

[4] K. Deshmukh et al., Chapter 4 - Mechanical analysis of polymers, Polymer Science and Innovative Applications, Elsevier, 2020, pages 117-152.

[5] Christian Bonten, 3 - Plastics Materials Engineering, Plastics Technology, Hanser, 2019, pages 63-246.

[6] Susan E.M. Selke et al., 4 - Major Plastics in Packaging, Plastics Packaging (Fourth Edition), Hanser, 2021, pages 105-164.

[7] TA Instruments, DMA Q800 specifications.

Determination of Shear Properties of 3D Printed Composite

P. Pavlíček [1, a], M. Fusek [1], M. Šofer [1]

[1] *Technical University of Ostrava*
[a] *pav0151@vsb.cz*

Abstract: Determination of mechanical properties in shear is important for defining the material model of composite. In this paper, the authors focus on the comparison of two types of shear experiment method when every method gave different result. From reason 3D is apparent non-homogenic structure of sample which have influence on mechanical property of composite. Standards of testing composite material do not take into account necessity removal of first layer of onyx. Understanding behavior of composite material under shear loading gave information about usability of tested method and for future research.

Keywords: 3D printing; Composites; Material model; DIC.

1 Introduction

This work is focused on methodology of testing 3D printed composites and obtain mechanical properties of composite under shear loading. For shear loading was written many standards but between these standards is significant differences between results and test procedure. In next case current standards neglect layers of material which must be removed. in 3D printed parts. These parts are more non-homogenic with comparison of traditional composite. This non-homogeneity of sample has significant dependency on mechanical property of composite Behavior under shear loading is important for define material model of composite and use inappropriate method has significant influence on interpretation of results.
Work is focused to comparation of these two experimental methods:

- ISO 527-4 [1],
- ASTM D5379 [2].

Understanding of behavior composite under shear loading is necessary for define suitability of testing method and for modification of standard for specifics of 3D printed composites. Accounting these influences into standards cause reduction of bad interpretation of results and raise of quality of measurement.

2 Material properties

Sample was printed on 3D printer Markforged X7 with accordance described above, material of sample is:

- Onyx composition of nylon and short carbon fibre,
- Carbon fibre.

Mechanical properties of used composite materials are written in tab. 1. Mechanical property of composite was measured with accordance ISO 527-4.

Tab. 1: Mechanical property of composite

Property	Value
Young modulus [GPa]	23.586
Yield stress [MPa]	194
Maximum stress [MPa]	714
Elongation [%]	4.21
Density $\left[\frac{g}{cm^{-3}}\right]$	1.4

3 Definition of Standards

3.1 Iso 524 – 4 Standard

Iso 527-4 is method of testing composites when is testing is done with unidirectional tension (compression). Shear testing in this method is done with orientation of fibers $\pm 45°$. Shape of sample is described in Fig.1 with indicated fibre orientation.

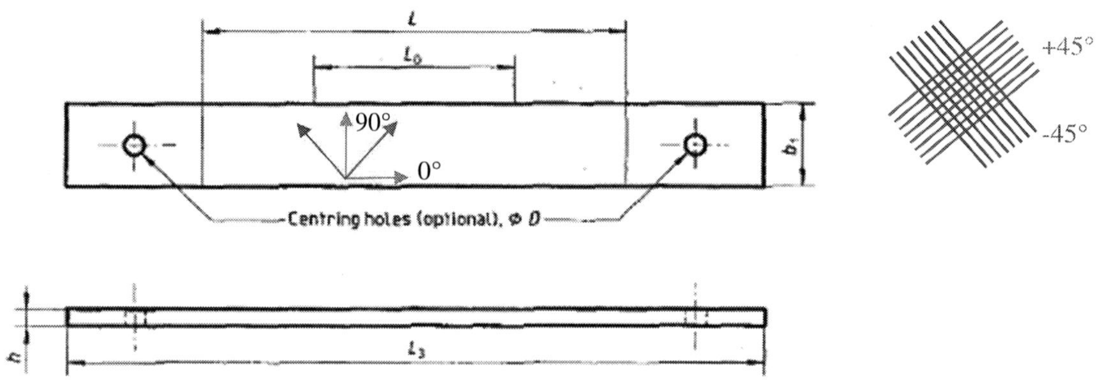

Fig. 1: Shape of sample and fibre orientation

Specimen shape is rectangular from reason even laying of fibers. In case of T-bone specimen, fibers are in the specimen deformed and this cause concentration in stress in transitional area. This standard defines equations for calculation mechanical properties when is important calculate stress, Shear modulus and strain. When stress was equated with formula (1).

$$\tau = \frac{F}{A} \tag{1}$$

When:

- F = force [N],
- A = area [mm^2],
- τ = shear stress [MPa].

When strain was equated with formula (2):

$$\gamma_{xy} = \frac{\epsilon_x - \epsilon_y}{2} \tag{2}$$

When:

- $\epsilon_{x(y)}$ =strain in fibre direction [−],
- γ_{xy} = shear strain [−].

Shear modulus was equated with formula (3):

$$G = \frac{\tau}{\gamma_{xy}} \quad (3)$$

When:

- G = shear modulus [MPa].

3.2 ASTM D5379 Standard

This method determines shear properties of composite materials reinforced with high modulus fibres. With this method can determined mechanical properties in all shear material planes. Specimen of ASTM D5379 standard is V-notched beam. which is represented in fig. 2 with shown fibres direction. On the left figure is measurement of strain on the specimen.

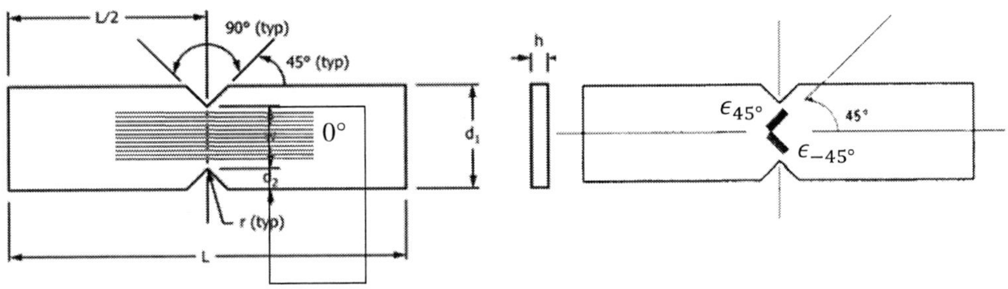

Fig 2: Specimen ASTM D5379 and laying of fibers

This method is more complex than ISO 527-4 for this method is need fixture for clamping and loading the specimen. Equations are same as in case ISO 527-4 but only with difference in indexes in equation (2).

3.3 Preparation of specimens

Specimen before test must be modified with use of milling machine when must be necessary remove surface layer of onyx. This layer has significant dependency on mechanical property of specimen. In the graph 1 is represented dependency on mechanical property of specimen.

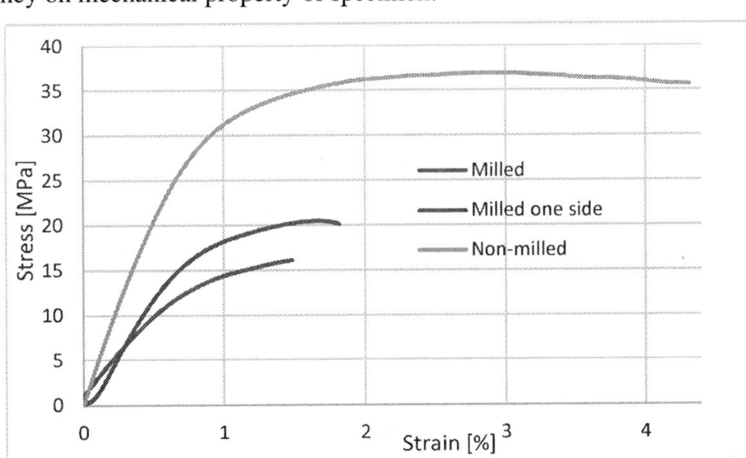

Fig 3: Behavior milled and non-milled sample

Milling of onyx layer has significant influence on mechanical property of sample mainly if the fiber in the sample has orientation 45°- 90 °. If the fibers are oriented with direction of loading effect of milling was neglected. Sample with fibers oriented in 90° direction has sample 2:25x times higher ultimate stress than milled sample. For correct testing of 3D printed composite materials is important remove onyx layers.

4 Homogeneity of composite

Homogeneity and structure of composite sample was measured with digital Microscope VHH 7000 with 200x magnification. On the figure 4 is represented homogeneity of composite sample. In the picture one is visible these findings:

ratio of fibres to matrix is 50 ± 10%%,

composite sample show inhomogeneous structure, with clusters of fibres when in the cluster have structure maximum fibre representation 80%,

inhomogeneous structure has significant effect on material parameters,

in zone out the cluster fibres has 20% representation,

diameter of single fiber is 4 nm,

composite layer does not contain onyx material.

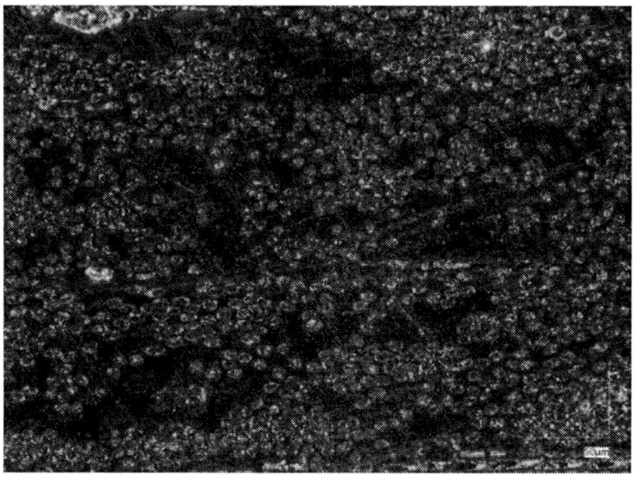

Fig 4: Sample homogeneity

Next investigation of homogeneity is focused on homogeneity of layering of fibres. Layering of fibres has influence of mechanical property of composite mainly if are fibres are not oriented in vector of loading. In 3D printing is presence of voids between fibres when voids are presented in fig.5.

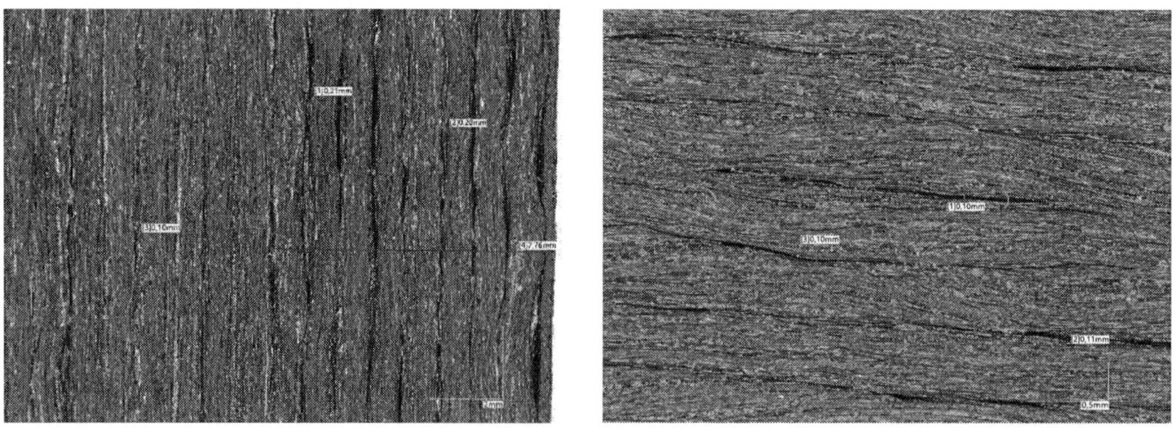

Fig. 5: Void presence

From the microscopy picture is visible presence of void when voids have size 0.1mm from the 7.75mm distance of the edge of sample. If are bellow this distance presence of void is higher than up to distance and size of void have value 0.2 mm. This zone is affected by 3D printing and has significant effect on mechanical properties of composite.

5 Test Procedure

Composites was tested on universal tensile machine Testometric MC-500 with 50kN force sensor. Test procedure was done with accordance ISO527-4 and ASTM D5379. On specimens before the test must be applied pattern with white chalk aerosol spray. Deformation of the sample was measured with DIC method with 3 frame per second. Specimen after clamping was preloaded with 100N. Test conditions are same for both methods, when test speed was 2 mm/min. On figures 6 and 7 are presented tested method.

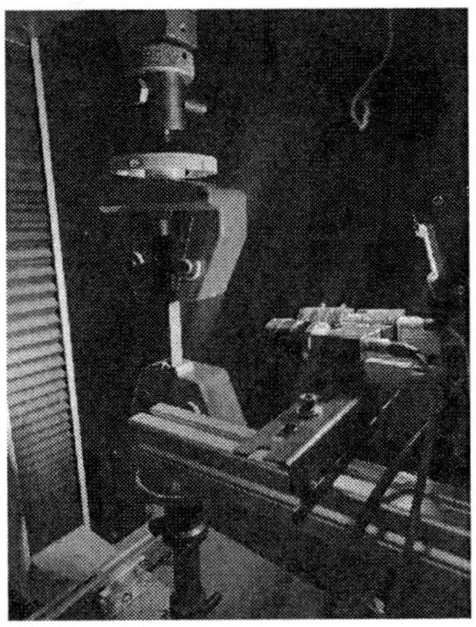

Fig. 6: ISO 527-4

Fig. 7: ASTM D5379

6 Results

From experiment was obtained test data which are written in tab. 2 and fig 8 and 9. Data was evaluated with accordance ISO 2602 [3]. From the test was visible dispersion of data in ISO 527-4 test. This dispersion was not visible in ASTM D5379 test.

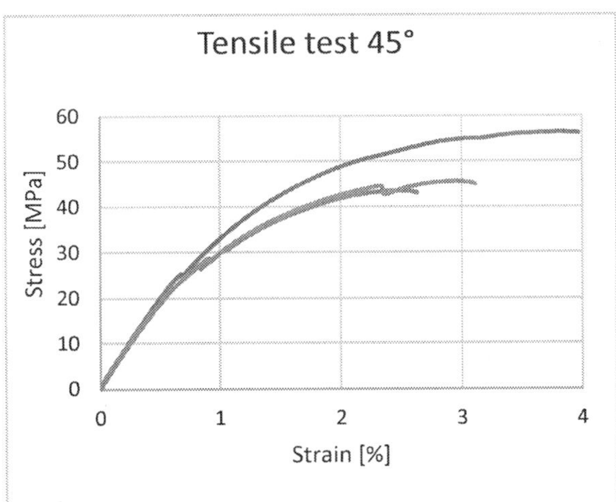

Fig. 8: ISO 527-4 Test

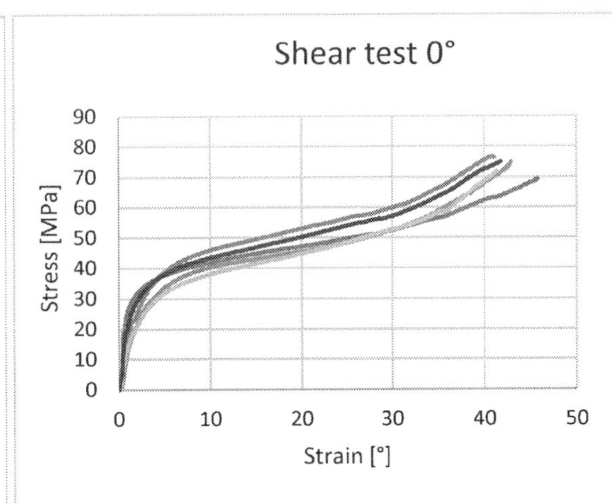

Fig. 9: ASTM D5379 Test

Tab. 2 Comparison mechanical properties

r	ISO 527-4	ASTM D5379	Difference [%]	Unit
Young modulus	3562 ± 142	2671 ± 172	35	MPa
Strength	48 ± 5.6	62 ± 4.4	29	MPa
Elongation	3.86 ± 0.8	36 ± 3.8	947	%
Yield Stress	26.6 ± 2.9	15.5 ± 2.2	42	MPa

From the results is visible difference in the graphs when ASTM D5379 gave higher results than ISO 527-4. These results are mainly visible in elongation when ASTM D5379 has different type of failure than ISO 527-4 test. Dispersion of data in ISO 527-4 test are caused by voids in the layer and with non-homogenic structure. Fracture in ISO 524-4 is mainly caused by interlaminar shear. In ASTM D 5379 has sample different fracture model when mainly fail fibers. From these results ISO 524-4 is more suitable for interlaminar shear test in composite than shear test.

7 Results of Fem Analysis

Fem analysis was done in Nastran 2022 and Marc 2022 model was modelled with shell Quad4 element with element size 1mm. Material model in analysis is orthotropic linear material. Element was check on mesh quality criteria. Model was evaluated through all layers with maximum value of stress.
Result of FEM analysis test ISO 524-4 is represented in fig.7.

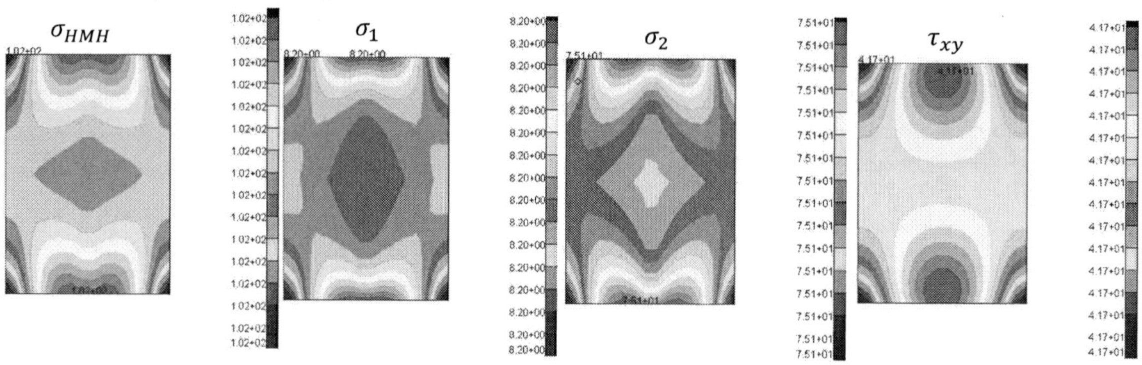

Fig. 7: Stress distribution ISO 524-4

From the fig. 7is visible stress in the center of the sample when maximum stress is distributed around center of sample. Shear distribution in the sample respond to simplified shear distribution. Results of FEM analysis from ASTM D5379 are represented in fig 8.

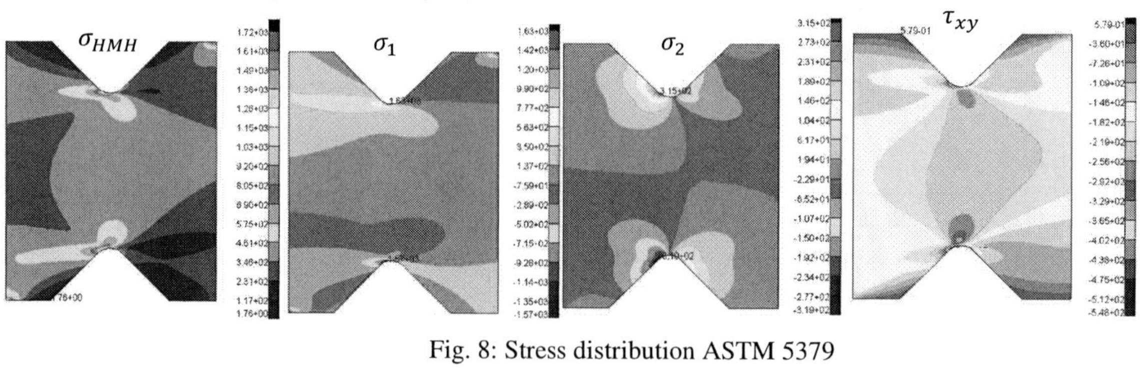

Fig. 8: Stress distribution ASTM 5379

From the fig 8 is visible distribution of stress when maximum stress is localized around radius of sample. Shear distribution in the sample respond to nonlinear shear distribution and is valid with shear theory. Secondary stress is cause by fibers bending and tension. This stress is valid with the test, From the both analysis is visible biaxial stress on sample when from the fem result is biaxial state is higher in ASTM D 5379 method.

8 Fracture under shear loading

From microscopy was found type of fracture from test ASTM 5379 when fracture is cause by shear and tension in the fibers. Detail of fracture is represented in fig. 16. From the picture are visible fractures of composite which are highlighted in red rectangular frame. 3D print in the root area has non-homogenic structure. This defect in homogeneity has significant effect on test and mechanical properties of material. Void between fibers is around 0.1 mm Failure of composite correspond to usual failure mode of CFRP composite under shear loading.

Fig. 16: Fracture under loading

9 Conclusion

From the results presented above was not found correlation between ASTM D5379 and ISO 527.4 both methods gave different results. Mean error between these methods is around 35%. Main difference between these two types of tests is in the different type of failure and stress distribution. 3D printing composite has:

- ratio matrix fiber around 50 ± 10%,
- 3D printed composite has non-homogenic structure,
- testing 3D printed composite without remove onyx layer has significant dependency on test results.

From the FEM results ASTM D5379 give more accurate results than ISO 527-4 method. ASTM D5379 is with compliance with shear theory. From the investigation ISO 527-4 is not recommended for testing composite under 45°, results from this test are only indicative and is usable for interlaminar shear test. ASTM D5379 give more accurate results and this method is recommended for shear testing. From the result of microscopy was found these results:

- print more likely desk and from the desk prepare samples, than printing samples,
- don't use composite 8 mm from the edge from higher occurrence of voids,
- don't print composite sample with complex shapes.

From the results of this work is necessary modified current standards for 3D printed composites, because don't reflect this type of technology. Concurrently if will not be considered assumptions of this work, results of the testing they will be error burdened. Future research of these problems is important for understanding of influences which influence measurement.

Acknowledgement

This work was supported by Specific Research „Experimental and Numerical Modelling for Solving Problems in Mechanics and Biomechanics "(SP2024/037).

References

[1] ISO 527-4. Plasty – Stanovení tahových vlastností – Část 4: Zkušební podmínky pro izotropní a ortotropní plastové kompozity vyztužené vlákny. Praha: Úřad pro technickou normalizaci, metrologii a státní zkušebnictví, 2023, 44 s. Třídící znak 640604.

[2] ASTM D5379. Standard Test Method for Shear Properties of Composite Materials by the V-Notched Beam Method, 13s. USA, Doi:10.1520/D5379_D5379M-19E01

[3] ISO 2602 Statistická interpretace výsledků zkoušek. Odhad průměru. Konfidenční interval, Praha: Úřad pro technickou normalizaci, metrologii a státní zkušebnictví, 1993, 12 s. Třídící znak 010231.

[4] ROHDE, S.; CANTRELL, J.; JEREZ, A.; KROESE, C.; DAMIANI, D. et al. Experimental Characterization of the Shear Properties of 3D–Printed ABS and Polycarbonate Parts. Online. Experimental Mechanics. 2018, roč. 58, č. 6, s. 871-884. ISSN 0014-4851. Dostupné z: https://doi.org/10.1007/s11340-017-0343-6. [cit. 2024-05-24].

[5] SAEED, Khalid. Characterization of 3D-printed laminate composites with enhanced mechanical properties. Online, D. Faculty of Computing, Engineering and Built Environment of Ulster University: Ulster University, 2022. Dostupné z: https://pure.ulster.ac.uk/ws/portalfiles/portal/105434798/2022SaeedPhD.pdf. [cit. 2024-05-24].

Characterisation of the Interfacial Transition Zone Between the Cement Matrix and Glass Powder Particles

Z. Prošek [1,2,a], A. Palička [1], P. Tesárek [1]

[1] Department of Mechanics, Faculty of Civil Engineering, CTU – Czech Technical University in Prague, Thákurova 7, 166 29 Prague, Czech Republic

[2] University Centre for Energy Efficient Buildings, CTU – Czech Technical University in Prague, Třinecká 1024, 273 43 Buštěhrad, Czech Republic

[a] zdenek.prosek@fsv.cvut.cz

Abstract: The article deals with determining of the width of the interfacial transition zone (ITZ) between the waste glass powder (WGP) and the cement matrix. The transition zone negatively affects the resulting mechanical properties of the composite. The article deals with surface treatments using plasma treatments under different gases to reduce the size of the ITZ, namely oxygen (O_2) and hydrogen (H_2) plasma. ITZ is measured using a scanning electron microscope (SEM) in combination with the determination of the percentage representation of individual elements using energy dispersive spectroscopy (EDS). The results are compared with mechanical tests on samples of 20 × 20 × 100 mm size. As part of the article, a reference sample without surface treatment was also measured. The results show that by using oxygen plasma it is possible to reduce the ITZ and thereby also improve the mechanical properties of the resulting cement composites.

Keywords: interfacial transition zone; waste glass particles; cement matrix; scanning electron microscope; mechanical properties.

1 Introduction

Cement composites are made up of three main components: matrix, aggregate, and interfacial transition zones (ITZ) [1]. Furthermore, the cement matrix consists of C-S-H gel, unhydrated clinkers, and portlandite (CH). The ITZ is acknowledged as the most vulnerable segment of cement composites, resulting in a reduced compressive strength compared to aggregate and hardened cement paste [2]. Furthermore, the presence of micro cracks within the ITZ contributes to changing the response of cement-based materials from linear to non-linear under loading conditions [3]. Mechanically, the extent of micro cracks within the ITZ is dictated by the degree of strain mismatch between the aggregate and the paste [4]. When stress is evenly distributed, differences in strain arise due to variations in the elastic modulus, the shear modulus of the aggregate, the matrix, ITZ, and the stress concentration. Serving as a bridge between the matrix and the aggregate, if the ITZ does not transmit stress effectively between phases, the compressive strength of the cement composite is compromised, even if one phase exhibits greater stiffness [5].

To understand the mechanism of formation of ITZ and its impact on the properties of cement composites, a detailed microstructural analysis was performed. It has been observed that the presence of CH, along with relatively high porosity, respectively, low-density C-S-H gel (LD CSH) and large capillary pores, significantly influences mechanical properties. Furthermore, the presence of large CH particles weakens van der Waals forces and has adverse effects on mechanical properties [6]. Apart from CH, micro cracks and high porosity are prone to forming within the ITZ, further diminishing mechanical properties. Another indicative factor for ITZ formation is the calcium-to-silicon (Ca/Si) ratio in the cement matrix. A ratio of 0.5-1.5 results in the formation of high-density C-S-H gel (HD CSH), while a ratio of 1.5 – 2.0 yields LD CSH [7]. Another significant effect is the high content of amorphous SiO_2 in WGP, exceeding 70 %, which, when exposed to an alkaline environment, can dissolve and generate a secondary C-S-H gel [8]. This pozzolanic reaction occurs gradually, enhancing the strength of the samples at later stages [9].

The focus of this article lies in the utilisation of WGP within cement paste, with the aim of enhancing its activation through surface plasma treatments. The objective is to decrease the thickness of the ITZ between WGP and the cement matrix, thereby leveraging the potential for the secondary formation of C-S-H gels. This approach aims to create a compact structure with a higher strength value compared to the reference sample.

2 Materials and Samples

The samples were composed of 80 wt. % Portland cement CEM I 42.5R (Radotín) and 20 wt. % of WGP, which comes from the waste material of the Recifa Packaging Industry, as. The water-to-binder ratio (w/b) was maintained consistently at 0.35. Furthermore, the glass powder underwent two different plasma treatments commonly employed to alter surface-wetting properties. The radiofrequency (RF) plasma treatment was conducted at low temperatures using a large-area, low-pressure system (AK 400, Roth & Rau). This process was carried out for 5 minutes in either oxygen or hydrogen atmospheres at a pressure of 15 Pa, with RF power set to 600 W and a substrate stage bias between 10 and 15 V, while keeping the temperature below 50 °C. This process, termed "room temperature" treatment, was repeated four times with periodic mixing of the glass powder to ensure uniformity.

From each batch, 5 samples containing various surface-treated WGP were created. These included a reference sample without surface-treated WGP (REF), WGP treated in an oxygen atmosphere (O_2), and WGP treated in a hydrogen atmosphere (H_2). The samples measured 20 × 20 × 100 mm and were destructively tested at the age of 3 months. Fragments from the destructive testing were embedded in epoxy resin under vacuum. They were then ground and polished in several steps to achieve a smooth surface. Grinding was done using SiC foils with grain sizes of 500 grains/cm², 1000 grains/cm², 2000 grains/cm², and 4000 grains/cm². For polishing, a suspension containing diamonds of 3 and 1 microns in size was used. Finally, the surface of the samples was coated with a conductive layer necessary for electron microscopy.

3 Experimental Methods

The samples were first tested for flexural strength using a three-point bending test. The support span during the test was 80 mm, and the loading was displacement-controlled at a rate of 0.5 mm/min. Subsequently, the broken beams were tested for compressive strength using a uniaxial compression test, which was also displacement-controlled at a rate of 3 mm/min. The effective loaded area was 2 × 2 cm. In both tests, the force at the time of sample failure was measured, which, in combination with the cross-sectional modulus or the cross-sectional area, provides information about the material's strength.

Another experimental method used was microscopic analysis. For this purpose, a ZEISS Merlin scanning electron microscope with a Schottky cathode was employed. The microscope is capable of detecting both secondary and backscattered electrons. In this study, backscattered electrons were analyzed as they provide information about phase contrast. The brighter the phase in the gray scale, the higher the atomic weight of the constituent elements. Additionally, an energy-dispersive spectrometer (EDS) from Oxford Instruments was used. Depending on the emitted X-ray energy, can directly determine the chemical composition of the examined area. To observe the transition zone, line scans were performed to monitor the mass content of calcium, silicon, and oxygen. The line scans ranged from 20 to 35 microns in length, with each scan consisting of 512 measurement points.

4 Results and Discussion

Results of the destructive tests indicate that oxygen plasma treatment improved both compressive strength and flexural tensile strength. The reference sample (REF) had a flexural tensile strength of 9.61 ± 0.30 MPa and a compressive strength of 84.9 ± 5.6 MPa. In comparison, the sample with a surface treated with oxygen plasma (O_2) exhibited a flexural tensile strength of 10.6 ± 0.59 MPa and a compressive strength of 90.0 ± 3.20 MPa. The samples treated with hydrogen plasma (H_2) showed the poorest performance, with a flexural tensile strength of 9.55 ± 0.12 MPa and a compressive strength of 78.3 ± 10.1 MPa. The results suggest that oxygen plasma positively affects the ITZ between the cement matrix and the WGP.

To confirm these findings, electron microscopy was conducted. In Fig. 1, line scans tested using EDS can be seen. According to the available literature, more than 20 different phases can form in the cement matrix, each with distinct structural and micromechanical properties. For simplification, the phases in the matrix can be categorized as follows: unhydrated clinker, which has a high atomic weight and appears white in the structure; portlandite ($Ca(OH)_2$), which appears light gray in images; C-S-H gel, which is gray; and pores and cracks, which are black. C-S-H gel is the most prevalent phase in the structure, and depending on the calcium-to-silicon ratio, it can be identified as low-density C-S-H gel (LD C-S-H) with poor mechanical properties or high-density C-S-H gel (HD C-S-H) with good mechanical properties. According to the literature, the Ca/Si ratio for LD C-S-H ranges from 1.5 to 2.0, and for HD C-S-H from 0.5 to 1.5 [7]. The line scan of the investigated area was always chosen in the region of C-S-H gel. The images in Fig. 1 show that for the reference sample (REF) and the H_2 sample, the size of the ITZ, where the Ca/Si ratio is higher, is 10 microns. In the case of the O_2 sample, the region is smaller, approximately 5 microns and a gradual increase in Ca/Si can be seen, indicating that there is no sudden change in mechanical properties at the interface between the cement matrix and the WGP. These results are consistent with the findings of the destructive tests.

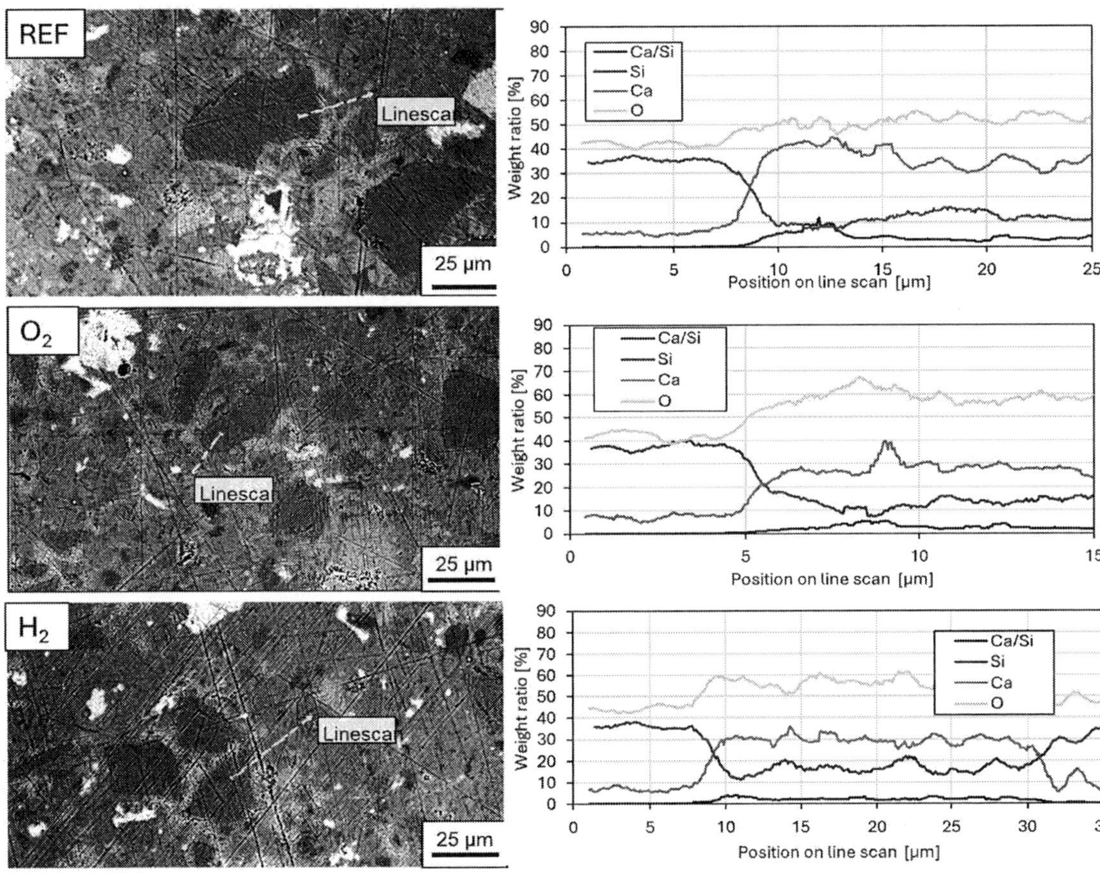

Fig. 1: SEM images of the backscattered electron detector magnified 500 × with the indicated area of line scan, evolution of the weight percentages of oxygen calcium and silicon in line scan.

5 Conclusion

In conclusion, oxygen plasma treatment significantly enhances the mechanical properties of cement composites with WGP. Destructive testing revealed that samples with the content of WGP treated with oxygen plasma showed improved flexural tensile strength and compressive strength compared to the reference and hydrogen-treated WGP. This improvement indicates a beneficial effect of oxygen plasma on the ITZ between the cement matrix and WGP.

Microscopic analysis supports these findings by showing a refined microstructure in the sample with oxygen-treated WGP. Scanning electron microscopy and energy-dispersive spectroscopy revealed a more gradual and smaller ITZ in samples with oxygen-treated WGP, approximately 5 microns, compared to 10 microns in the reference and hydrogen-treated WGP. The smoother transition zone correlates with enhanced mechanical properties, confirming that oxygen plasma treatment effectively improves the interface quality and material performance in cement composites with treated WGP.

Acknowledgement

This paper was financially supported by the GA CR research project under the number GA23-05500S and by Czech Technical University in Prague under No. SGS project SGS24/037/OHK1/1T/11. The authors also thank Recifa a.s.

References

[1] A. M. Neville, Properties of Concrete, London, Longman, 1995.

[2] Y. Ke et al., Identification of microstructural characteristics in lightweight aggregate concretes by micromechanical modelling including the interfacial transition zone (ITZ), Cement and Concrete Research 40 (2010) 1590–1600, https://doi.org/10.1016/j.cemconres.2010.07.001.

[3] T. Akçaoğlu, T. Mustafa, T. Çelik, Effect of coarse aggregate size and matrix quality on ITZ and failure behavior of concrete under uniaxial compression, Cement and concrete composites 26 (2004) 633–638, https://doi.org/10.1016/S0958-9465(03)00092-1.

[4] G. Prokopski, B. Langier, Effect of water/cement ratio and silica fume addition on the fracture toughness and morphology of fractured surfaces of gravel concretes, Cement and Concrete Research 30 (2000) 1427–1433, https://doi.org/10.1016/S0008-8846(00)00332-X.

[5] J. A. Larbi, The cement paste-aggregate interfacial zone in concrete, Delft, TU Delft doctoral thesis, 1991.

[6] N. Van Tuan et al., Hydration and microstructure of ultra high performance concrete incorporating rice husk ash, Cement and Concrete Research 41 (2011) 1104–1111, https://doi.org/10.1016/j.cemconres.2011.06.009.

[7] Z. Prošek, V. Nežerka, P. Tesárek, Enhancing cementitious pastes with waste marble sludge, Construction and Building Materials 255 (2020) 119372, https://doi.org/10.1016/j.conbuildmat.2020.119372.

[8] Y. Kong et al., SEM analysis of the interfacial transition zone between cement-glass powder paste and aggregate of mortar under microwave curing, Materials 9 (2016) 733, https://doi.org/10.3390/ma9090733.

[9] H. Du, K. H. Tan, Concrete with recycled glass as fine aggregates, ACI Materials Journal 111 (2014) 47–58, http://dx.doi.org/10.14359/51686446.

Bending Analysis of GFRP Composite Plate with a Hole Using Carbon Fiber Sensors

N. Schmidová [1,a], K. Doubrava [1], Z. Padovec [1], D. Blaha [1], R. Kottner [2], C. Novotný[1], M. Růžička [1]

[1] *Division of Strength and Elasticity, Faculty of Mechanical Engineering, Czech Technical University in Prague, Technická 4, 166 07, Prague, Czech Republic*
[2] *ACO Industries Tabor s.r.o., Průmyslová 1158, 391 01, Sezimovo Ústí, Czech Republic*
[a] *Nikola.Schmidova@cvut.cz*

Abstract: Glass fiber composite vessels are used for underground water storage and surface water treatment. Especially during the installation of such vessels damage occurs. This paper investigates possibility of utilization of integrated Carbon Fiber Sensors (CFSs) for Structural Health Monitoring (SHM) of such glass fiber composite vessels. In this study, a flexural analysis of a glass fibre-reinforced composite panel with a central hole was performed. The strain monitoring using CFSs near the hole was done. The obtained results confirm the possibility of using CFSs for SHM of GFRP vessels.

Keywords: Structural Health Monitoring; Carbon Fiber Sensor; Glass Fiber Composite vessel.

1 Introduction

A system of drainage channels and retention tanks is usually used for the collection and eventual subsequent treatment of surface stormwater from large, paved areas, allowing the water management cycle to be implemented from water capture, through treatment to final retention or recycling.

The cylindrical glass fiber composite wound vessels of interest are used to collect these waters and e.g. separate the oil from water. Vessels of various volumes are placed usually underground vertically or horizontally. Increased loads and stresses on the vessel walls, particularly localized loads, may occur during the placement of the vessels in the backfill and during actual operation, which may cause damage to the integrity of the vessel. It would therefore be desirable to develop a system for the possible indication of such local stresses, or of disruption of the vessel composite structure itself.

One of the critical areas of the vessel is the area around the mouth of the GFRP vessel. Sensors made of carbon fiber tow (Carbon Fiber Sensors - CFSs) integrated during manufacturing process can be used for structural monitoring of the vessel.

An initial measurement using CFSs were performed during bending test of the plate type specimen with a central hole.

2 Methods

2.1 Specimen Preparation

The specimens were manufactured using an automated filament winding procedure of glass fibre tows with polyester resin. The length of the specimen was 400 mm, width was 200 mm and thickness 9.76 mm. The central hole had a diameter of 80 mm. The specimens were cut out from a wound vessel. Total three specimens were manufactured for the measurement using CFS.

The CFSs were manufactured according to procedure published in [1] in a length of 80 mm. The CFSs were attached to the surface of the specimen after manufacturing and covered with a covering layer, see Fig. 1.

The three CFSs were placed along the longitudinal axes of the specimen according to Fig.1. The CFS sensors were positioned in such a way, that one sensor was installed closest to the hole. The second sensor was placed in the distance of 20 mm from the hole and the third sensor was installed in the distance of 40 mm from the hole. Total three CFSs were installed on each of two specimens 1 and 2. On specimen 3, 6 CFSs were installed, each three symmetrically on both sides of the hole (sensor group A, sensor group B).

Along with the CFSs was installed also optical fiber on each of the specimens. The optical fiber was used as a Distributed Fiber Optic Sensing (DFOS) system. Results from DFOS system will be described only briefly in this paper. More details can be found in Blaha et al. [2].

2.2 Sensors for Monitoring

The CFSs in this application are used for strain measurement, but they are meant to be used also for crack and impact monitoring in future as published in [1] and [3]. The change in measured electrical resistance of CFSs is recorded during loading. The change of measured electrical resistance is proportional to the extension of the sensor. Each sensor gives an integral information about its extension, which could be used for calculation of average strain along the sensor.

Measurements of the changes in electrical resistance change were performed using a Quantumx MX 1615B strain gauge amplifier, manufactured by Hottinger Baldwin Messtechnik GmbH. The output of the Wheatstone bridge was used to determine the changes in the electrical resistance of the measured carbon fiber sensors.

The change of measured electrical resistance of each CF sensor was converted to average mechanical strain using formula given in [3]. The strain sensitivity k of a CFS equal to 1.75 was used.

The second type of sensor, the DFOS sensors, enables measuring strain with high spatial resolution. It is used as a reference method. The method is based on the principle of Rayleigh scattering. Luna Odisi 6100 series was used for data recording from DFOS sensor.

a) prepared sample with optic and carbon fiber sensors installed on the plate

b) installation of covering glass mat layer on top of the sample

Fig. 1: Preparation of the experimental samples

2.3 Experimental Procedure

The bending test of the GFRP plate with a central hole was performed in the configuration depicted in Fig. 1. On the Heckert FPZ100/1 universal testing machine. The configuration of the performed 4-point bending test was as follows. The distance between lower supports was 320 mm, the distance between upper supports was 120 mm. The experimental procedure comprises of 5 loading cycles. The loading force varied from 0 to 1 kN.

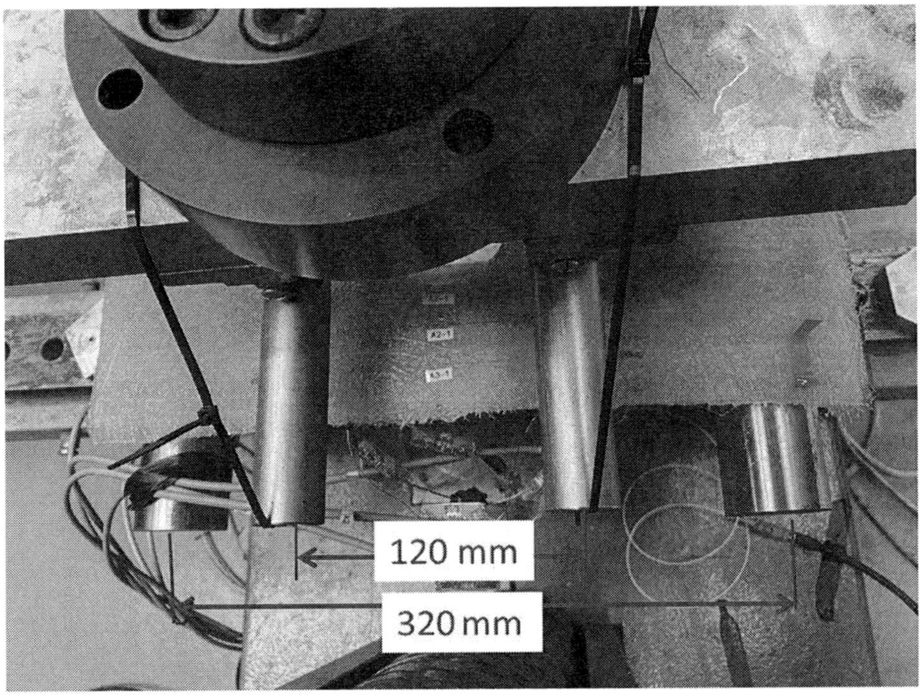

Fig. 2: Configuration of the bending test of the specimen with a central hole and integrated Carbon Fiber Sensors.

2.4 Analysis using Finite Element Method (FEM)

Numerical simulation of the 4PB text was performed using Abaqus software. The model of the specimen was prepared as a shell model according to specification of the specimens. The material properties were set according to used material (glass fibers, polyester resin). The data of mechanical strain was read from the FE model along the path, where sensors were placed, see Fig. 3.

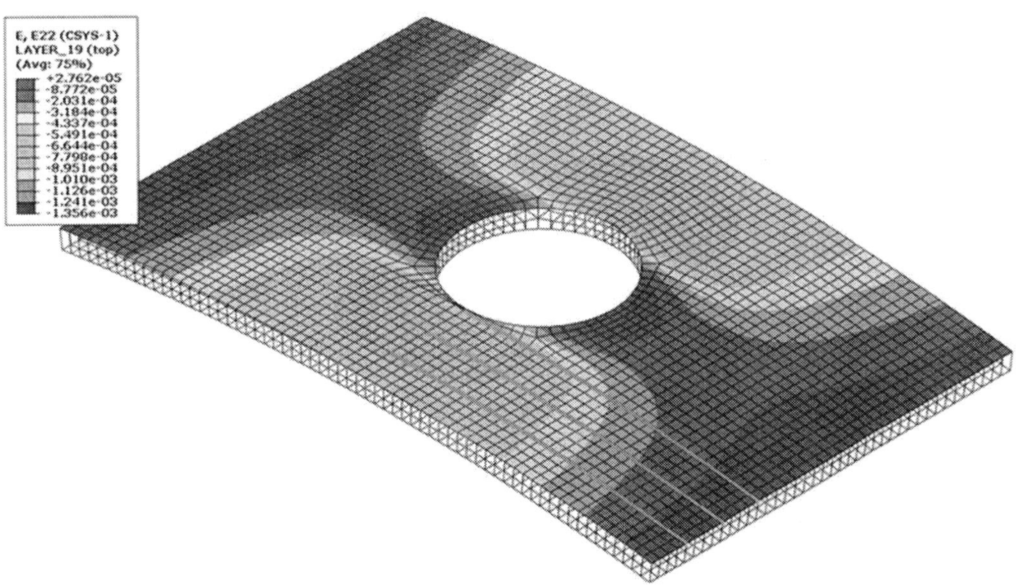

Fig. 3: Results of FE simulation of 4PB test.

3 Results and Discussion

The loading comprises of five loading cycles, measured data are given in Fig. 4.

Specimen 3 was loaded two times of 5 cycles, in order to measure DFOS sensors on both sides of the specimens, so for this specimen data for loading cycles 6 to 10 will be presented.

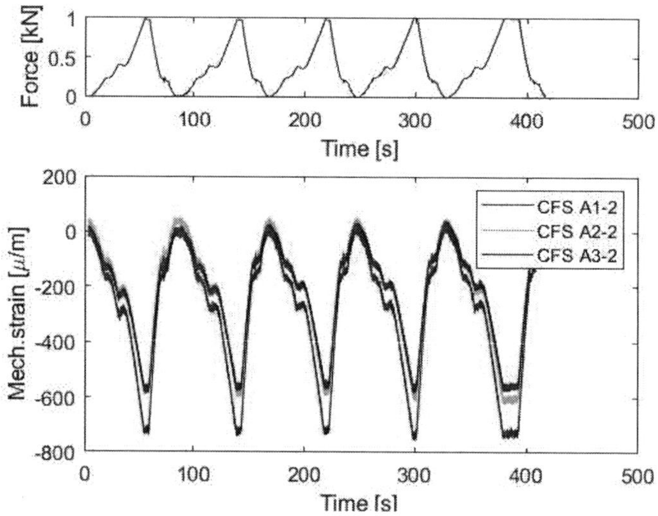

Fig. 4: Record of measured strain using CF sensors and applied force during 5 cycles of loading of specimen 2.

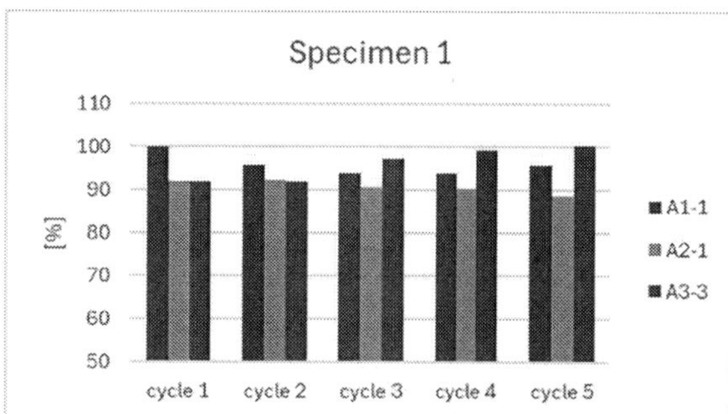

a) percentage change of measured strain on specimen 1

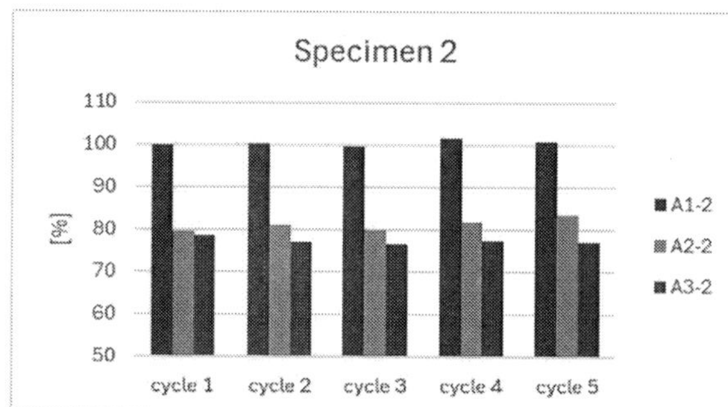

b) percentage change of measured strain on specimen 2

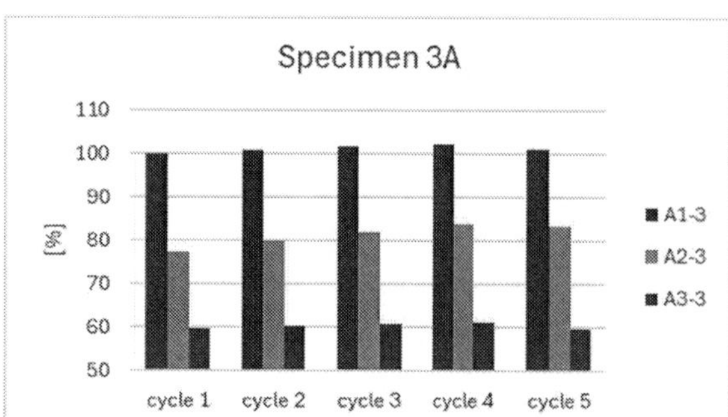

c) percentage change of measured strain on specimen 3 – group of sensors A

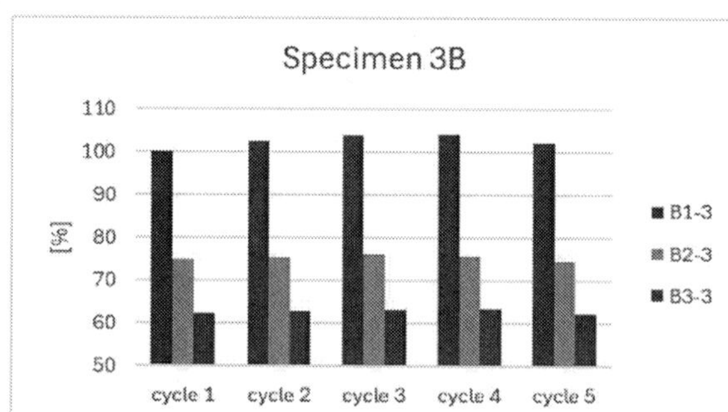

d) percentage change of measured strain on specimen 3 – group of sensors B

Fig. 5: Percentage change of measured strain using CF sensors on different specimens on different locations during five loading cycles.

Tab. 1: Comparison of measured data from CF and DFOS sensors and FEM simulation.

		Specimen 1		Specimen 2		Specimen 3A		Specimen 3B		Average all specimens	
Sensor location	**FEM [%]**	CFS [%]	OF av. [%]	CFS [%]	OF av. [%]	CFS [%]	OF av. [%]	CFS [%]	OF av. [%]	**CFS [%]**	**OF av. [%]**
Path 0	**100**	79	85	98	100	118	128	112	115	**102**	**107**
Path 20	**83**	75	80	79	89	98	-	86	92	**84**	**87**
Path 40	**74**	79	81	75	81	69	-	71	79	**74**	**80**

In Fig. 5 are given data of measured from CF sensors as relative values for better comparison. Measured strain for first cycle of sensor closest to the hole corresponds to 100 %. Change in measured data with the number of cycles is relatively small for sensors to all specimen with the exception of CF sensor A3 installed on specimen 1.

Measured data from CF sensors were compared to data gained from DFOS sensors and simulation using FEM. From each CF sensor only one value of average strain along the length of the sensor was obtained. Therefore, average strain value from measured strain along DFOS sensors was also calculated. Mechanical strain in longitudinal direction along paths corresponding to sensor localization was also read from the FEM analysis and average strain value was determined.

The data obtained from the simulation were compared with the measurement data from DFOS and CF sensors. Results of such comparison is given in Tab. 1. For better comparison the data are given as a percentage values, where 100 % corresponds to strain along sensor A1 (path 0) in FEM simulation.

In Fig. 6, Fig. 7 and Fig. 8 are compared data from CF and DFOS sensors and also FEM simulation for specimen 2.

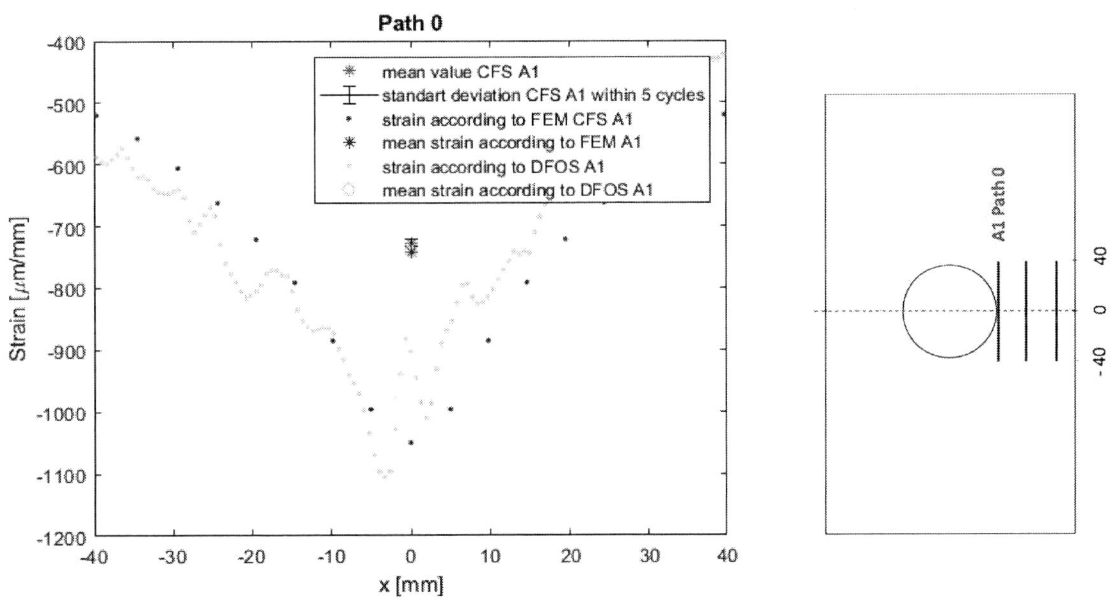

a) mechanical strain along path 0 at specimen 2 b) location of sensor A1

Fig. 6: Comparison of results for Specimen 2 along path 0 corresponding to CF sensor A1, DFOS sensor and FEM simulation.

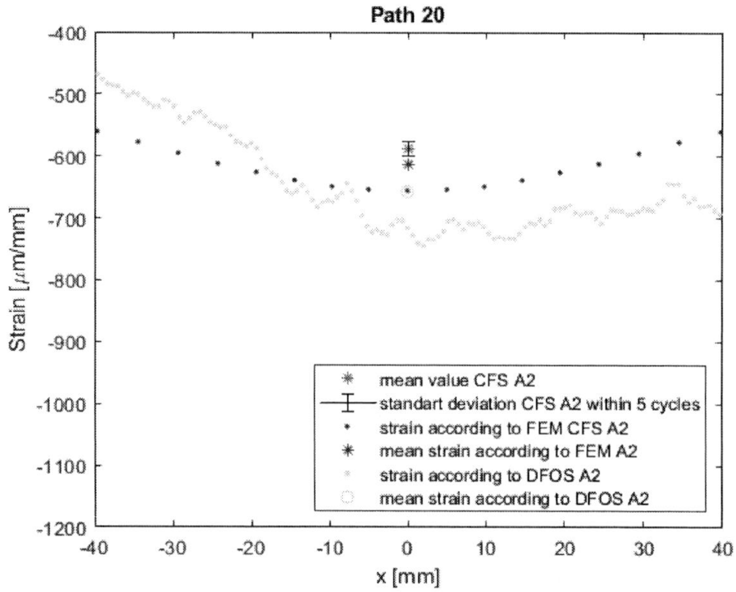

a) mechanical strain along path 20 at specimen 2

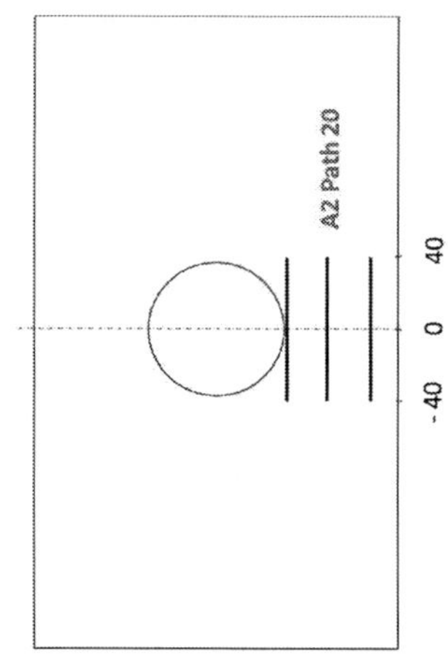

b) location of sensor A2

Fig. 7: Comparison of results for Specimen 2 along path 20 corresponding to CF sensor A2, DFOS sensor and FEM simulation.

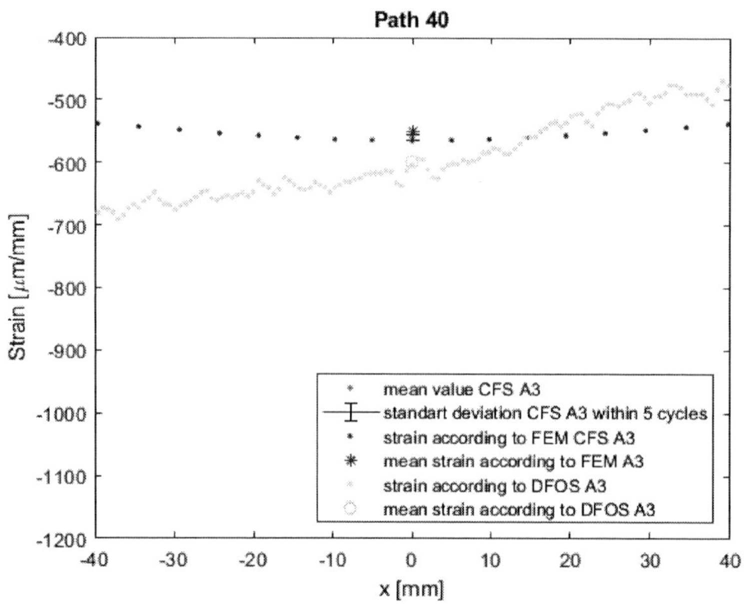

a) mechanical strain along path 40 at specimen 2

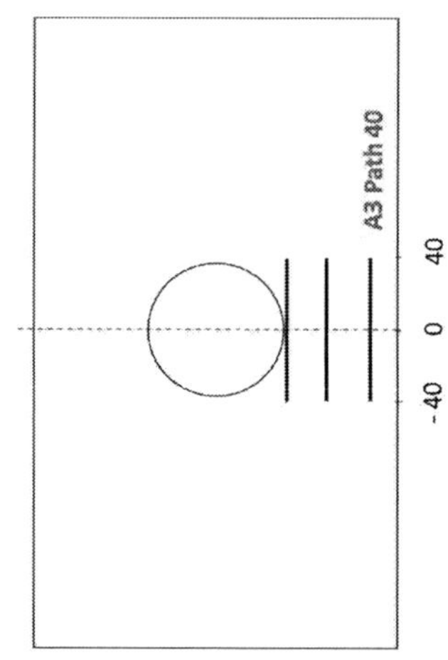

b) location of sensor A3

Fig. 8: Comparison of results for Specimen 2 along path 20 corresponding to CF sensor A2, DFOS sensor and FEM simulation.

4 Conclusion

The experimental campaign brought initial information about measurement strain on filament wound GFRP composite plate during repeated loading near the hole. Measured data were compared among CF sensors installed on different locations within one specimen and among several different specimens.

The results obtained from the strain measurements using CFSs show good repeatability and are in good agreement with the expected strain distribution based on the theory and FEM analysis. Measured average strain by CF sensors are in good agreement with the measured data from reference DFOS sensors.

The advantage of CF sensors and optical fibers is that they can be directly integrated into the composite structure of the vessel. The advantage of the usage of CF sensors over DFOS sensors is the significantly lower price of the measurement device, although we get only one average value of strain over the whole length of the sensor.

Acknowledgement

The project TN02000010 of National Competence Centre of Mechatronics and Smart Technologies for Mechanical Engineering is co-financed from the state budget by the Technology Agency of the Czech Republic within the National Centres of Competence Programme.

References

[1] A. Horoschenkoff, Ch. Christner, Carbon Fibre Sensor: Theory and Application, in: Composites and Their Applications, ed. N. Hu, Intech Open, [online], 2012, DOI: 10.5772/50504.

[2] D. Blaha, Z. Padovec, K. Doubrava, R. Kottner, C. Novotný, N. Schmidová, M. Růžička, Bending Analysis of Glass Fiber Reinforced Composite Using Distributed Fiber Optic Sensing System, in proc: Experimental Stress Analysis 2024, 62th International Conference (2024), Research and Testing Institute Pilsen, Boží Dar, Czech Republic.

[3] Th. Matzies, C. Christner, A. Horoschenkoff, H. Rapp, Carbon fibre sensor meshes: Simulation and experiment, Computational Materials Science 64 (2012) 122–125, DOI: 10.1016/j.commatsci.2012.05.070.

Validation of Simulation of the Temperature Field of a Part during Selective Laser Melting Procedure

R. Skácelík [1,a] , M. Španiel [1]

[1] *Department of Mechanics, Biomechanics and Mechatronics, Faculty of Mechanical Engineering, Czech Technical University in Prague, Technická 4, 160 00 Prague 6, Czech Republic*

[a] *radek.skacelik@fs.cvut.cz*

Abstract: The heat that is supplied to the printed part during the selective melting procedure must be dissipated so that the part does not overheat. The procedure fulfilling this requirement can be effectively designed using validated numerical simulations. The measurement of the temperature field on the top layer of the part using a thermocamera placed in the area of printing is the main topic of the contribution. The measured temperature field is compared with one obtained from simulation employing the finite element method. The comparison provides us with the validation of both the finite element model and computational procedure which are described in the paper as well.

Keywords: 3D printing; thermal camera; finale element method; recoating; selective laser melting.

1 Introduction

Selective laser melting is one of the additive technologies based on melting metal powder. The principle consists of consecutive application of thin layers of metal powder in a workspace that is locally melted at the sites of the printed part. This creates a solid occupying the entire geometry of the printed part while the rest of the workspace remains filled with powder [1]. The processing parameters, i.e. laser input power, scanning speed and slit pitch, have a significant effect on the heat supplied and, consequently on the temperature of the part. This heat is dissipated through the part to the platform. The part is subjected to dimensional changes due to the thermal expansion of the material. This reduces the quality and the consistency of the thickness of the new layers applied. In extreme cases, this can result in the closure of the gap between the part and the coating device, causing print failure. Such accidents reduce the efficiency of the entire process and therefore it is an effort to be able to predict them before the actual printing of the part takes place.

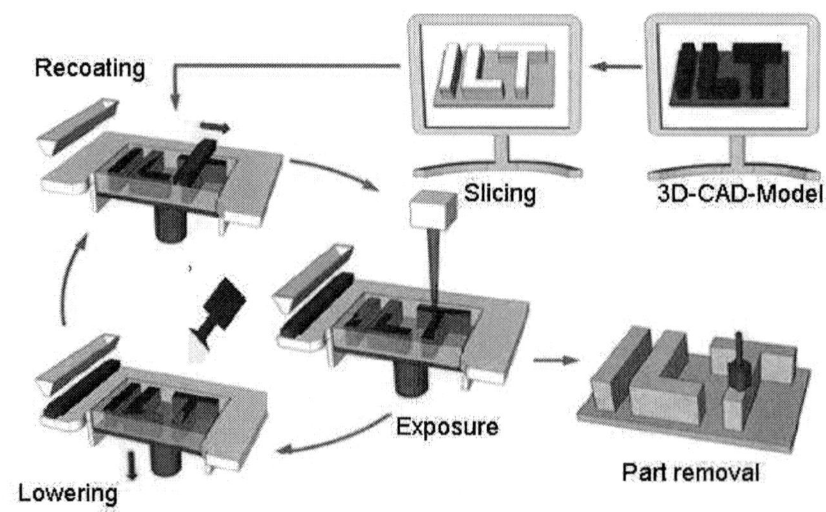

Fig. 1: Printing scheme with the location of the measurements in the printing cycle indicated. [3]

The motivation of this work is to validate a numerical model to describe the distribution of temperature fields inside a printed part. Due to the limited possibilities to verify the temperature distribution of the part, a thermal camera was used to measure the residual temperature of the sample (Fig. 1, lowering section). This was used to measure the temperature of the surface before the coating was swiped. These temperatures are monitored to observe the history and trends of temperature evolution as the geometry of the part changes.

2 Methods

2.1 Printer Configuration

In this work, the TRIUMF 1000 printer was used for the measurements. This printer has a circular print base with a diameter of 100 mm and the height of the print area is approximately 150 mm. A thermal camera (Fig. 2) was installed in the press compartment, and during the creation of the new layer, the camera lens was covered by a servo-controlled aperture. The servo was controlled depending on the position of the coating and the printing time. The measurements were performed with a FLIR Lepton 3.5 thermal imaging camera whose compact dimensions allow it to be placed close to the printing area. The pixel density of the measurement is about 1 pixel per square millimeter. The raw data was transferred from the thermal camera directly to the computer where the data was processed and averaged. The sample was placed in the middle of the platform. The measurement point was chosen in the middle of the sample and the temperature values were averaged using the surrounding pixel values.

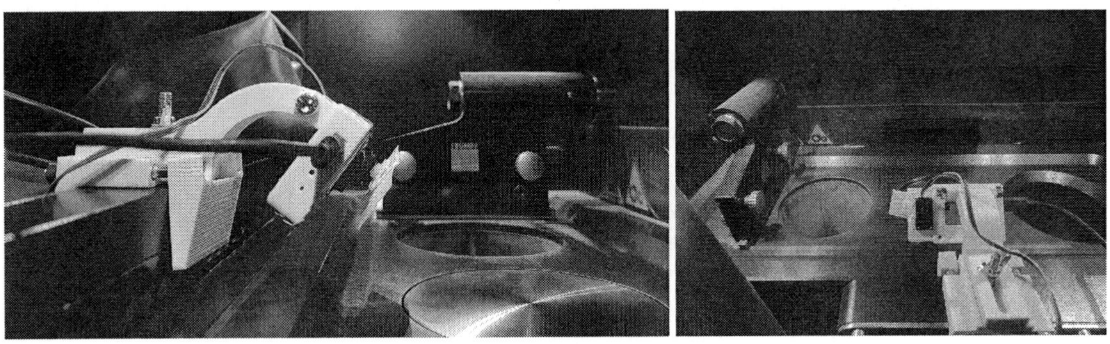

Fig. 2: Printer area configuration.

2.2 Calibration of the Thermal Camera

The thermal camera was calibrated using a modified method described in [1]. The model was not heated directly in the printer but was heated in an annealing furnace. Subsequently, the sample was fitted with a thermocouple. The thermocouple was placed in a groove printed directly into the sample so that it had the largest possible heat transfer surface near the top of the cube (Fig. 3). After the thermocouple had equilibrated the temperature with the sample, the emissivity coefficient of the surface of the top of the cube was adjusted according to this temperature value so that the temperature value from the thermocouple matched that from the thermal imager. The sample was then scanned throughout the cooling period to determine the deviation between the thermocouple and thermal camera values.

Fig. 3: Sample on the platform with thermocouple.

2.3 Experimental Sample and Build Plan

Measurements were taken on a sample (Fig. 4) that has a circular base and then widended at a 45-degree angle. The rotationally symmetric body was chosen to eliminate as much as possible the effect of asymmetric cooling due to heat transfer from the body to the powder. MS-1 tool steel was chosen as the specimen material. The laser power is 250 W, the scanning speed is 1000 mm/s, the grating is 0.1 mm and the layer thickness is 0.04 mm. The coater waited 4 s after each layer was applied before the next layer was applied. It took 3 s to apply a new layer of powder. The frequency of the laser return to a single point is shown in Fig. 5. In this case, the amplitude correlates with the area of each layer of the body. This is due to the location of a single sample on the platform. The sample is deliberately chosen so that the total amount of heat delivered to the sample increases as a function of the area of the top layer.

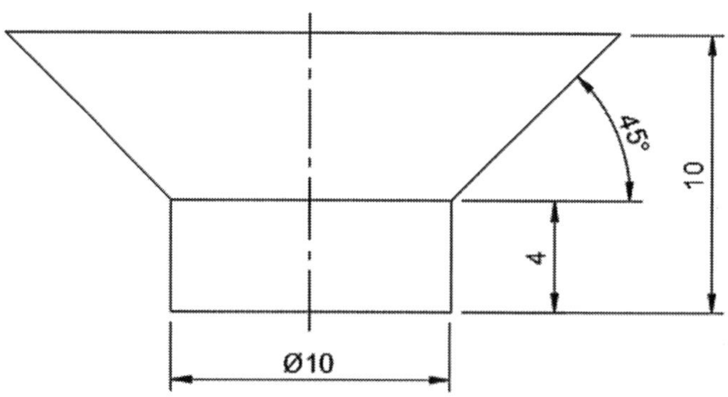

Fig. 4: Geometry of the sample.

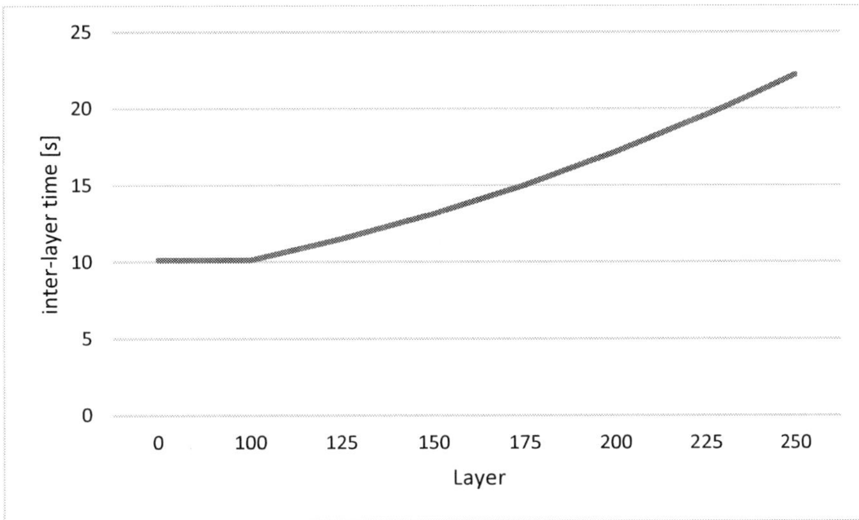

Fig 5: Internal layer time of the printed sample.

2.4 Numerical Simulation

The same parts as in the experiment were also used to calculate the temperature fields by simulation of the printing procedure using the finite element method. The mesh density and intensity of the temperature source were adjusted so there would be a match with the experiment. A modified Oofem and Code-Aster solver code was used for the calculation. The sample model was meshed using a voxel-type mesh (Fig. 6). This mesh includes a portion of the base.

Fig. 6: Mesh of the sample (brown) and base (purple).

The simulation was performed using the flash heating method. This method Is based on heating the entire top layer at once. The hot top layer serves as a loading boundary condition from which the heat is then distributed to the entire model. In this type of simulation, the temperature of the added layer is set as the solidification temperature of the molten metal. The cooling time of the part before the next layer is applied is determined from the internal layer time (ILT) (Fig. 5). In the case where multiple layers (super layer thickness SLT) are applied simultaneously, the cooling time is increased in proportion to the number of layers merged.

3 Results and Discussion

By comparing the temperature of the sample measured by the thermocouple and the thermocamera, the emissivity coefficient of the printed part surface was found. This coefficient was determined to be 0.35 for the tool steel used.

By measuring the temperature history of the sample, it was found that the part heats up with increasing top layer area (Fig. 7), despite the fact that the ILT increases with increasing area (Fig. 5).

The finite element method was used to predict the temperature history of the sample. Several different element sizes (SLT) were used and the result was compared with the temperature waveform of the experimental sample. Sufficient agreement with the experiment was achieved for elements with a thickness of 3 real layers (Fig. 7).

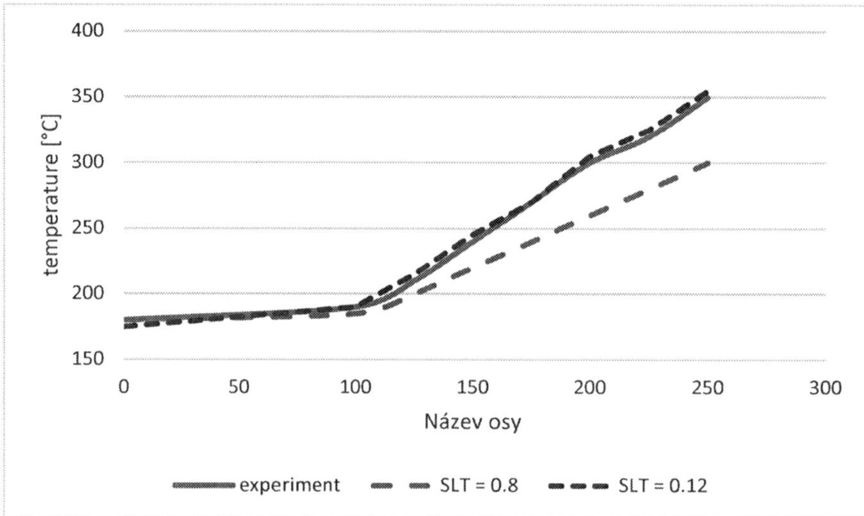

Fig. 7: Results of simulation and measurements.

4 Conclusion

Calibration and temperature measurements were made using a thermal imager during printing at the moment before the new layer was applied. These values were used to validate the results of the finite element simulation.

References

[1] T. DebRoy, H.L. Wei, J.S. Zuback, T. Mukherjee, J.W. Elmer, J.O. Milewski, A.M. Beese, A. Wilson-Heid, A. De, W. Zhang, Additive manufacturing of metallic components – process, structure and properties, Progress in Material Science 92 (2018) 112–224, https://doi.org/10.1016/j.pmatsci.2017.10.001.

[2] M. Montazeri, P. Rao, Sensor-based build condition monitoring in laser powder bed fusion additive manufacturing process using a spectral graph theoretic approach, Journal of Manufacturing Science and Engineering 140 (9) (2018) MANU-18-1023, https://doi.org/10.1115/1.4040264.

[3] US Research Nanomaterial, "Strain less steel powder" [online], available at: http://www.us-nano.com/inc/sdetail/14838. [Zugriff am 15 September 2017].

Methodologies for Modeling and Testing Bus Hybrid Structure Roll-Over Dynamics

S. Špirk [1,a], T. Kalina [1], J. Václavík [2]

[1] *University of West Bohemia, Czech Republic*
[2] *research and testing institute Pilsen, Czech Republic*
[a] *spirks@fst.zcu.cz*

Abstract: The realm of bus construction is a multifaceted and intricate domain, characterized by diverse requirements. Among these, the mass of the body frame stands out as a crucial parameter. A lighter body frame often translates to greater occupancy potential, yet it typically compromises the structural rigidity. However, striking a balance between reduced mass and maintained stiffness is achievable through geometric optimization or the integration of cutting-edge materials. This paper delves into a novel structural concept for bus construction, incorporating carbon fiber layers and foam fillings as localized reinforcements throughout the framework. The primary objective is to explore the viability of these hybrid joints within the bus construction landscape, encompassing hybrid beams and structural nodes.[3]

Keywords: bus, hybrid, composite, foam, crash.

1 Introduction

The primary stress on the bus framework predominantly arises from fatigue and collision scenarios. While steel design offers lower raw material costs compared to composite alternatives, the manufacturing time for steel designs significantly exceeds that of composites.[1] This study hones in on crash dynamics, particularly focusing on roll-over scenarios (Ro.66), where composite materials and manufacturing techniques play a pivotal role in reducing vehicle weight for buses and trams.[2] Efforts to streamline bus construction, while maintaining or enhancing stiffness, unfold across two fronts. Firstly, establishing a numerical model, grounded in available material data, serves as the initial framework for experimental endeavors. Secondly, validating this numerical model entails cross-referencing it with empirical measurements. A prudent approach involves validating individual materials independently, followed by more complex assemblies, such as steel beams integrated with composite strips and foam. Though the core of the work revolves around a simplified beam model, the overarching findings will directly inform advancements in bus construction methodologies. More information about the details of the numerical model is provided in an older post [6] that focuses exclusively on validation using three-point bending.

2 Numerical model of a bus segment

The car body segment model was created based on CAD data provided by the manufacturer of the specific bus. As the main software the explicit solver PAM-CRASH. Similar results can be achieved with another dominant explicit solver.[4] The primary simulated part is the load-bearing structure model. Additionally, wheel models were created with a relatively coarse mesh to define the position of the platform and its pivot point. Although the wheels are not essential for the calculation and are deleted in some versions of the calculation, they are crucial for determining the initial state of the platform and creating the surface into which the car body segment impacts. The angle and speed at which the segment impacts are derived from free fall and were calculated using simple analytical relationships, so the simulation can start just before the impact.

The finite element mesh is mainly composed of quadrilateral elements (with one Gauss integration point) and only occasionally isolated degenerate triangular elements (node3 = node4). The characteristic size of the elements is 5 mm, similar to the three-point bending test simulation, ensuring compatibility of the material models used (element size dependence). The meshes are as regular as possible to prevent the occurrence of "phantom" stress waves. The detail of significant mesh areas is shown in the following images (Fig. 1,2).

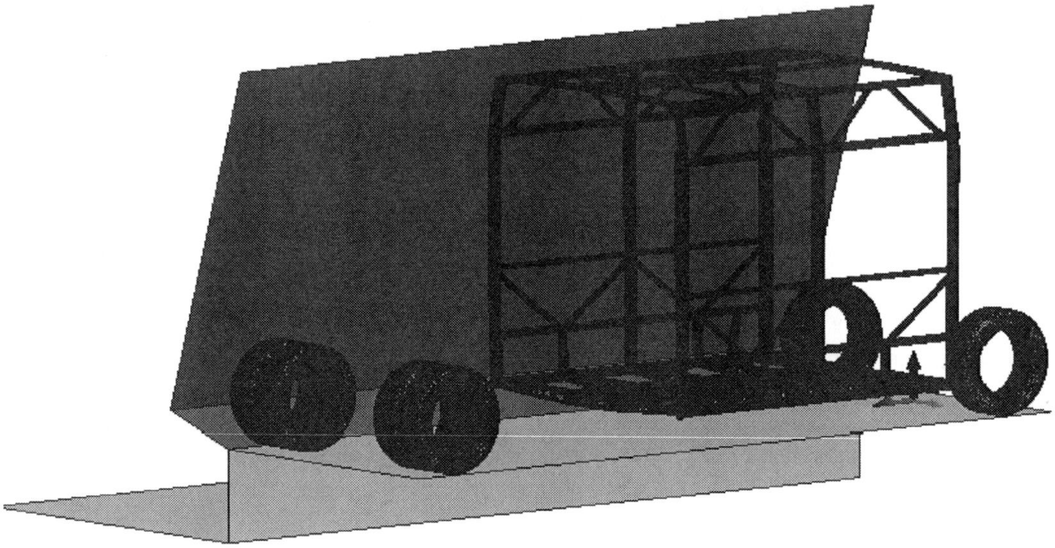

Fig. 1: The numerical model of a bus segment

Welds in the model are idealized with a rigid "Tide" connection. This simplified model does not consider weld stiffness or the possibility of its failure, but elements surrounding the joint partially compensate for this function. This idealization is sufficient for the case in question, and there is no need to use a "Plink" connection method or to model the welds with volume elements. In the model, a single contact is defined between the corner of the segment and the surface it impacts. It is a symmetric node-segment "33" type contact, where the coefficient of friction was set at 0.1[-].If necessary, the model can be supplemented with self-contact (type 36) in areas of significant plastic deformation. The floor into which the segment impacts is created with very coarse elements and serves only to define a rigid body. The center of mass of this body has all degrees of freedom removed. The mass of the segment is increased by non-structural masses. This mass represents the additional equipment of the bus that is not modeled in the simulation as its contribution to stiffness is negligible. The mass increase was made considering the mass of the entire bus and the ratio of the segment length to the length of the whole bus.

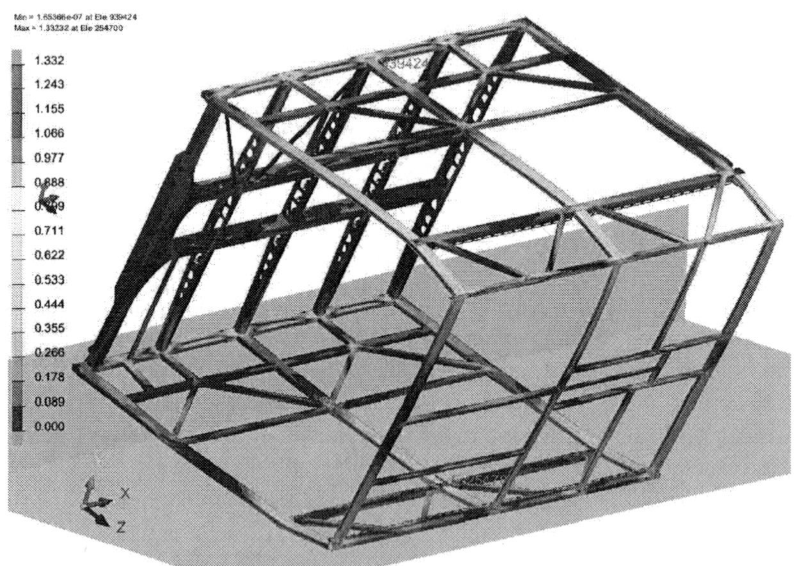

Fig. 2: The Segment during roll-over test Von Mises Stress [GPa] (80 ms)

3 Numerical model of a ring-type sample

Based on the discussion of experimental possibilities and representativeness, the manufacturer proposed a "RING" type sample. This sample represents the smallest possible section of the bus body segment and is suitable for static or quasi-static tests, the results of which will serve well for validating the numerical model. It should be noted that such a narrow segment does not meet the standard requirements, and this was anticipated in advance. On the other hand, the approval process can be supported by numerical calculation. A segment consisting of two rings (to meet the standard requirements) would pose a risk of uneven loading during the collapse of one of the beams. Therefore, such a sample may not be entirely suitable for validating the numerical model (considering the relatively unconventional reinforcement methods). A sample with at least three posts is already relatively large, and its testing would be considerably more demanding. Furthermore, as the previous numerical calculation shows, even such a part may not accurately represent the behavior of the entire bus body. An undeniable advantage of the "RING" type sample is also the possibility of conducting many more tests, which allows, for example, testing the structure with and without CFRP strips, or obtaining more detailed results for a single post (in the segment, only the average deformation characteristic for all posts would be determined).

The following image shows the geometry of the sample. The discussed reinforcements are visible, as well as the indentations on the inner part of the segment, which are (according to the manufacturer) necessary to create space for the electrical installation. Since it is not possible to show the entire mesh due to its fineness and the dimensions of the geometry, the following images show at least the details of the finite element mesh (Fig. 3). The mesh is created with similar parameters as mentioned in the previous chapter. It is a relatively regular mesh with a characteristic element size of 5 mm. The welds are idealized similarly to the previous case using a rigid "Tide" connection. It is a node-segment connection where a constant mutual distance is maintained during the calculation by a penalty algorithm. A preview of the individual welds is shown in the following image (Fig. 4).

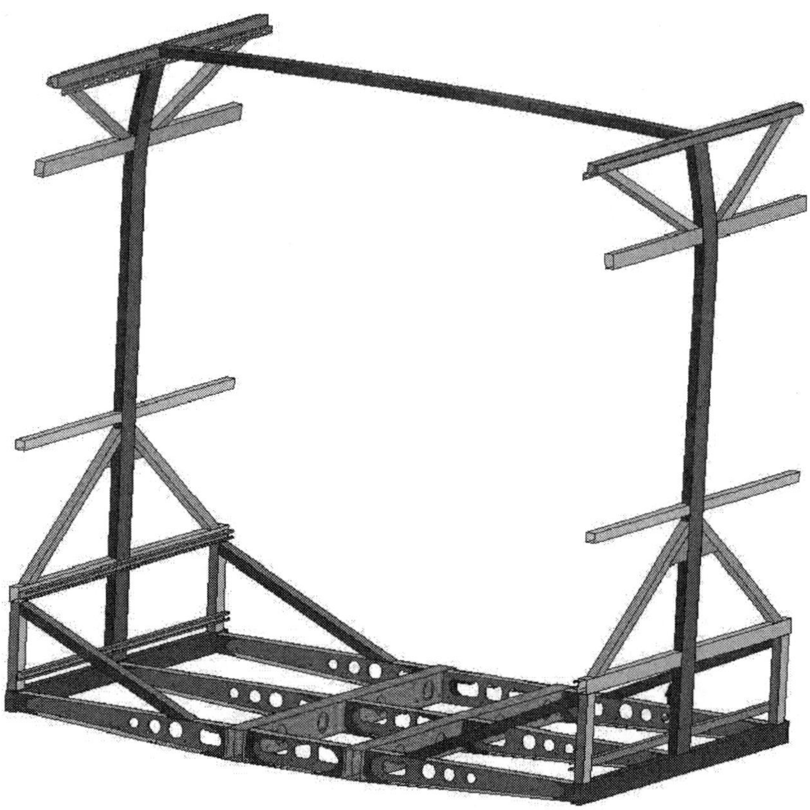

Fig. 3: The numerical model of sample named as “RING”

4 Proposal of a Test Procedure Supported by Numerical Simulation

Due to better contact conditions, the possibility of transferring forces to the structure using a generally planar element with sufficient rigidity was discussed and simulated. Loading by a plane requires significantly complicated mounting to enforce displacement, so the displacement of the sheet was gradually enforced again using a rope.

Since, unlike the real car body, the corresponding rotation of the car body (occurring during a rollover impact) cannot be simply ensured in the case of a simplified test, there is unrealistic contact of the surface with the structure in the lower part of the sheet. The structure essentially rests on the sheet and is shaped by it from the point of contact to the bottom edge of the sheet. Initially, the sheet was gradually shortened for this purpose, so that in the end, a cutout for the beam was made in it. This prevents unrealistic straightening of the beam by the sheet.

Fig. 4: The loading method of the sample demonstrates the testing procedure

5 Simulation results

The simulation results serve as important support for the experiment. Some significant results are shown in the images below. These primarily include the lateral deformation of the structure, the field of plastic stresses in the most exposed parts of the structure, and the indication of the angle by which the vertical beam bends due to plastic deformations. The image below clearly shows the overall deformation of the "ring" type sample. Two critical points are clearly visible in the overall structure where structural stability collapse occurs and where the largest local plastic deformations appear. These are the vertical main beam, which collapses at the point above the diagonal bracing, and the horizontal roof beam, which collapses at the corner of the roof structure. For better illustration, a polygon representing the crew's survival space was added in the numerical simulation. The polygon is doubled because the bus has two levels of floor height and thus two levels of passenger seats.

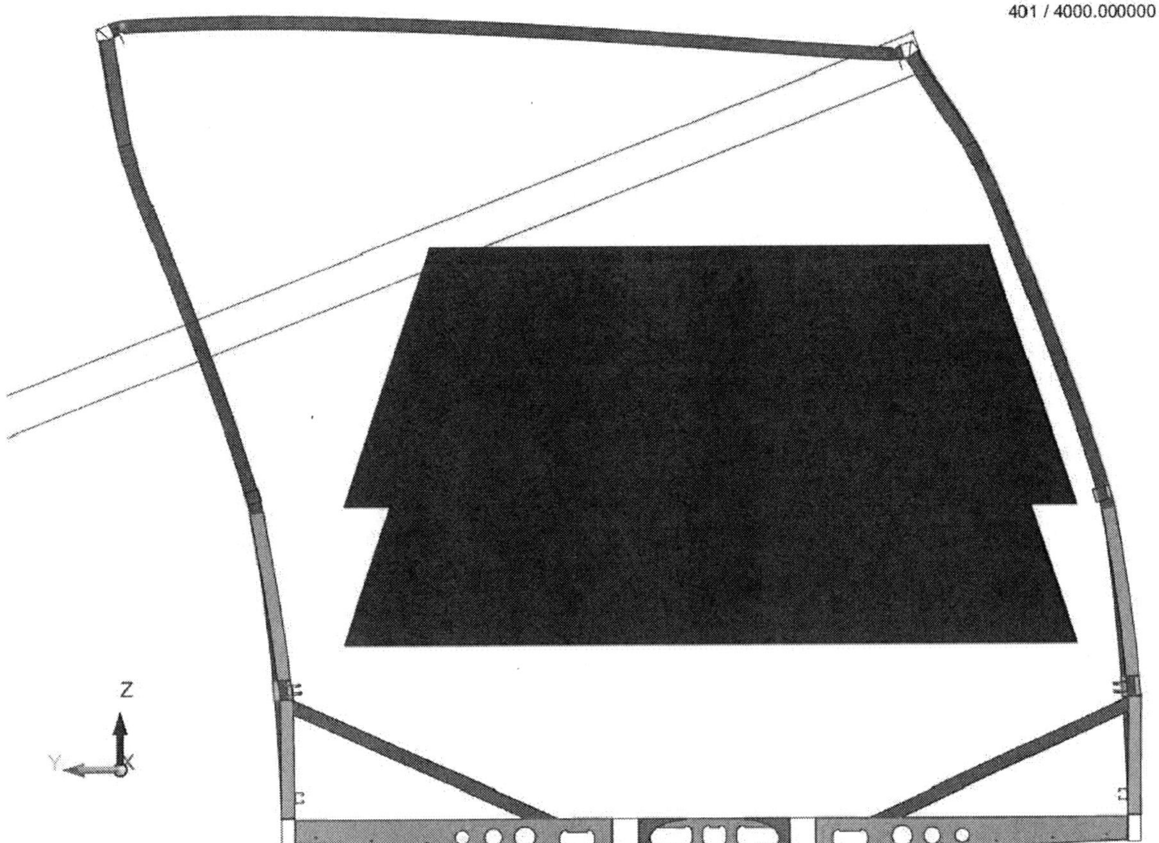

Fig. 5: The deformation of whole structure with illustrated survival space, time 4000 [ms]

The two images below very precisely illustrate the plastic deformations at the two critical points described above. In the top image, the collapse of the structure at the indentation is clearly visible. This geometric element (according to the manufacturer) is used for cable routing. It is evident that the area where the diagonal braces end will always appear as the weakest point. The behavior is similar to a cantilever beam. The indentation itself significantly influences the manner and occurrence of plastic deformations. However, during the experiment, the plastic deformations at this point were somewhat smaller. This can be explained by a certain geometric discrepancy, where the provided CAD data corresponded more to embossing, while the transverse section of the physical profile exhibited a more V-shaped curve. By partially adjusting the geometry of the numerical model, at least a somewhat better match was achieved.

In the bottom image, significant plastic deformations in the roof corner are visible. Neither during the simulation nor during the experiment did the structure fail at this point. It is clear that this part represents a "bottleneck" significantly affecting the strength of the entire structure. Unfortunately, no reinforcement is possible here due to surrounding equipment and installed cabling. CFRP strips placed on both the top and bottom sides of the roof significantly improve the overall condition. However, it is a major drawback that composite reinforcements generally drive plastic deformations increasingly into the corner, which is not desirable.

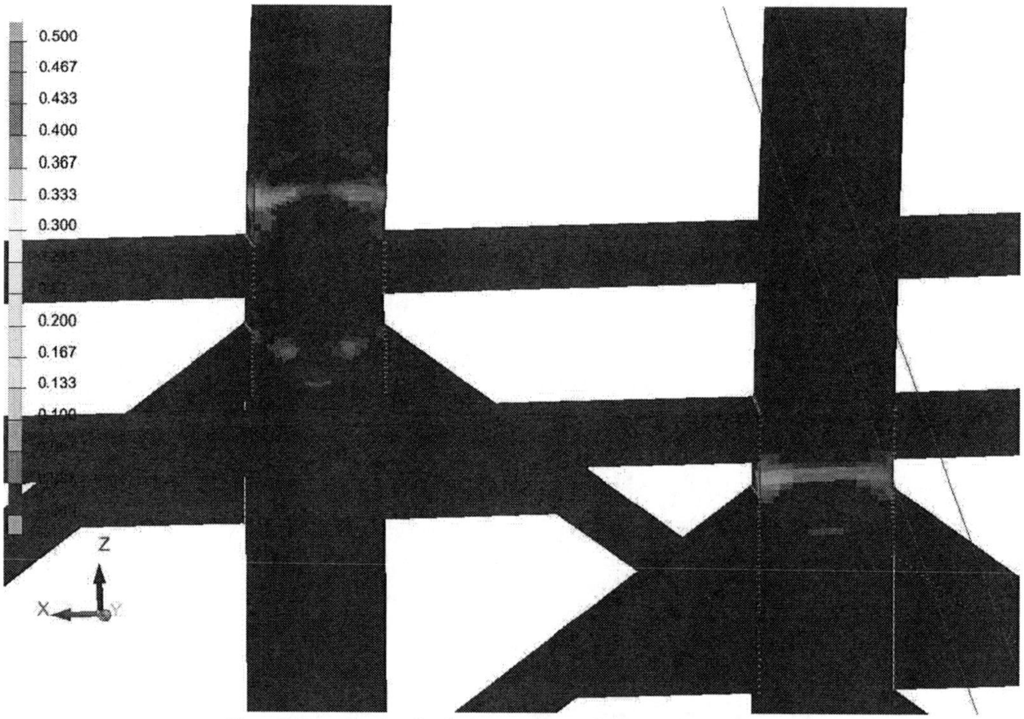

Fig. 6: Local plastic deformation of the main beam [-]

Fig. 7: The local plastic deformation in critical corner of the roof [-]

6 Results discussion

Probably the most significant results are summarized in the graph in the image below. The graph demonstrates a fairly good agreement between the results of numerical simulations and experiments. The numerical simulation achieves the most favorable result, where both the outer profiles (on the outer sides) and the horizontal profile of the roof (on the top and bottom sides) are reinforced. This was expected. A less expected result is the relatively small benefit of increasing stiffness for the foam-filled profile, both for a simple steel profile filled with foam and for a combination of reinforcement with foam and composites. A very interesting result from the numerical simulation is the finding that the reinforcement of the upper strip has a dominant influence on the roof. This reinforcement certainly significantly contributes to the overall increase in stiffness in the variant with the maximum number of reinforcements. The contribution of the reinforcement of the side profiles is very close to the contribution of the reinforcement of the roof horizontal beam. These results were achieved despite being suboptimal and without using more detailed optimization methods as described in the article [5].

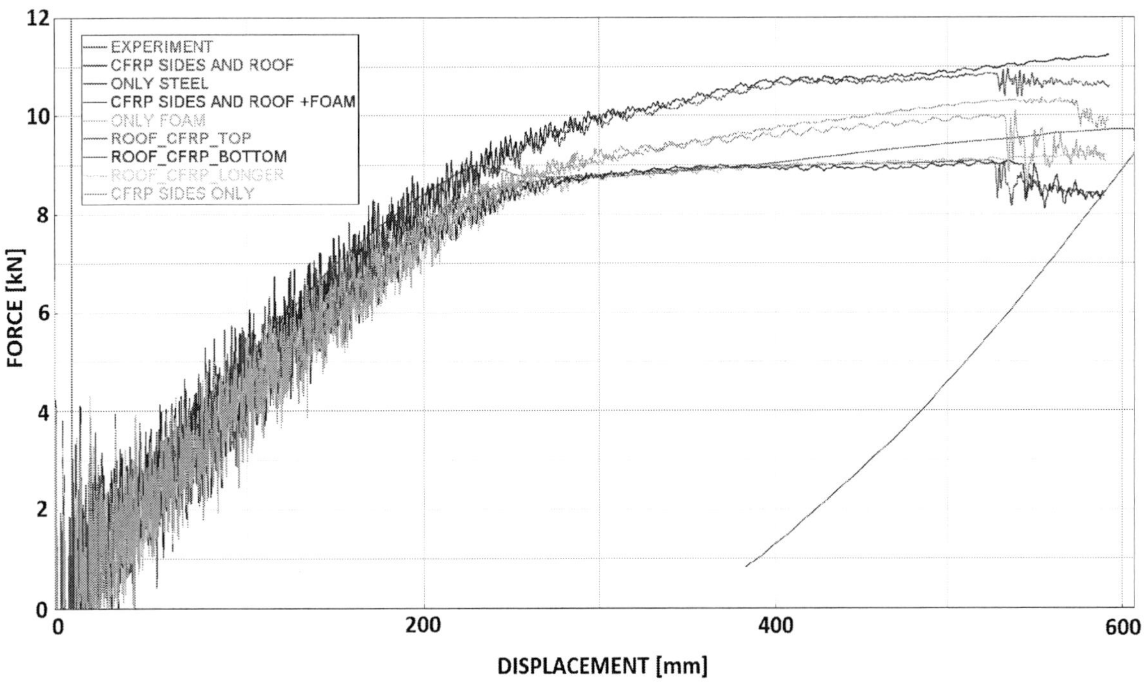

Fig. 8: The deformation characteristic of all simulated variants and its comparison with experiment

7 Conclusion

The paper briefly describes the numerical simulations created in the field of passive safety. First, the simulation of a bus segment impact on a rigid surface is presented. This involves a rollover test of the segment. A relatively low stiffness was found in the corners of the structure, which led to the modification of the "RING" type sample. The "RING" type sample represents the smallest possible section of the bus segment and is ideal for validating the numerical model. This sample was physically manufactured and is expected to be tested.

To support the conducted tests, several calculations representing different methods of substitute loading were created. Since static loading of the structure with a force at an angle (corresponding to the load during a bus rollover) is assumed, various possibilities of transferring the force to the structure were primarily addressed without artificially increasing its stiffness. Loading with an inclined plate would be relatively simple (and its simulation was very easy); however, this loading method seems to be practically difficult to implement. On the other hand, a relatively simpler method might be loading using a steel rope. Within these calculations, a variant with a 15° lateral deflection of the rope was also simulated. The results of this calculation do not indicate any danger of stability loss (unintended deflection of the structure in the direction of the vehicle's movement).

The paper also presents preliminary findings and methodologies for numerically simulating hybrid beam structures. Initial insights are gleaned from a partially validated model featuring a steel hollow profile, augmented with a composite strip and foam filling. Observations indicate that optimal performance is achieved when the composite material closely matches the Young´s modulus of steel, highlighting the importance of effective collaboration between composite and steel components. The integrity of the glue connection emerges as a critical factor in ensuring the structural robustness. Validation of material models can be augmented through three-point bending tests and their corresponding simulations. However, for comprehensive validation, it is recommended to conduct numerical simulations alongside full-scale experiments on bus segments. While such experiments are slated for the future, they promise to offer invaluable insights into the performance and behavior of the hybrid beam structure under real-world conditions.

Acknowledgement

This work has been supported by the project TH71020003 "Composite reinforcement in a light stainless steel bus structure" provided by the Ministry of Industry and Trade of Czech Republic.

References

[1] E. Larodé at al. A new concept of a bus structure made of composite materials by using continuous transversal frames. Composite Structures 32 (1995), 345-356.

[2] K. Hee-Young at al. A study on the crashworthiness and rollover characteristics of low-floor bus made of sandwich composites. Journal of Mechanical Science and Technology 23 (2009), 2686-2693.

[3] M. Kriescher at al. Possibilities for the use of metal-hybrid-structures for vehicle crash load cases. Proceedings o 1. International Conference Euro Hybrid Materials and Structures 2014, 10-11. April 2014, Stade, Germany.

[4] E. Acar, B. Yilmaz, M.A. Güler, et al. Multi-fidelity crashworthiness optimization of a bus bumper system under frontal impact. J Braz. Soc. Mech. Sci. Eng. (2020), 42, 493.

[5] P. Kunakorn-ong, K. Ruangjirakit, P. Jongpradist, S. Aimmanee, Y. Laoonual. Design and optimization of electric bus monocoque structure consisting of composite materials. Proceedings of the Institution of Mechanical Engineers, Part C: Journal of Mechanical Engineering Science. (2020), 234(20):4069-4086.

[6] S. Špirk, J. Václavík. New Structure Concept of the Bus Construction with the Addition Carbon Fiber Layers and Filling Foams. EAN 2021 - 59th International Scientific Conference on Experimental Stress Analysis - Book of Full Papers, pp. 218 – 221.

Biaxial Cruciform Testing of Fiber Composite Materials

V. Vomáčko[1,2,*], M. Šulc[2], J. Kolář[2]

[1] *Faculty of Mechatronics, Informatics and Interdisciplinary Studies, Technical University of Liberec, Studentská 1402/2, 46117, Liberec, Czech Republic*

[2] *Department of Virtual and Applied Mechanics, VÚTS, a.s., Svárovská 619, 46001, Liberec, Czech Republic*

* *vaclav.vomacko@tul.cz*

Abstract: The biaxial cruciform testing method was adopted to validate the failure criteria results obtained through numerical simulations. A biaxial testing machine was developed, and carbon fiber-epoxy laminate specimens were tested. Failure strains were evaluated in the central area of the specimens. Specimens without tabs failed in the arm section, while specimens with glass-fiber tabs failed in the central section. However, both failed prematurely, as confirmed by numerical simulations with a progressive damage model. The first ply failures for the matrix and fibers were evaluated using simulations. The values of the first ply failure strain in the simulation were much lower than the measured ultimate failure strains. This demonstrates that the use of biaxial tests can be advantageous for determining the ultimate failure of components in combination with simulations, even in cases where the specimens fail prematurely.

Keywords: biaxial; cruciform; composite; failure; validation.

1 Introduction

The design process of a composite structural part comprises several stages, as depicted in Fig. 1. Initially, conceptual ideas are sketched, and the concept of the part is developed. This is followed by the part's design using CAD software. Numerical simulations play a pivotal role in optimizing the component and conducting failure analysis. The accuracy of the numerical model heavily relies on precise input data. However, not all material parameters can be directly measured, leading to the need for determination through various means. For instance, to fully define the stiffness matrix of an orthotropic linearly elastic material, nine constants are required, including three Young's moduli, three shear moduli, and three Poisson's ratios. Additionally, determining the material's strengths is critical for accurate failure analysis. This process typically involves a combination of mechanical tests, micromechanical models and educated guesses, which introduces uncertainty into the outcomes of numerical models. To simulate fracture behavior, a progressive damage model can be utilized. Furthermore, in the case of thin laminates, a plane stress simplification is applied, reducing the number of independent material parameters to five [3].

Various failure theories (Max stress. Max strain, Hoffman, Tsai-Wu, Hashin, Puck, LaRC04, etc.) can be used to evaluate failure based on computed stresses and strains in simulations [1]. Each failure criterion predicts failure in a distinct manner. That is the reason why failure prediction of composites is not so straightforward as in the case of isotropic materials.

Different approaches can be employed to validate safety limits of parts from fiber composites obtained from numerical simulations. Uniaxial mechanical test data alone may not be adequate for conducting reliable failure analysis under complex stress conditions [4]. Traditionally, validation of simulations involves prototyping and performing experiments to simulate real-world conditions. Alternatively, biaxial tests can be employed. Two common methods are used to conduct biaxial tests on fiber-reinforced composites. The first involves applying a combination of axial (tensile/compressive), torsional, and internal or external pressure loading to tubular specimens, which creates a biaxial stress state. Tubular specimens were utilized in the World Wide Failure Exercise [5]. The second approach uses planar cruciform specimens, where the desired biaxial stress state is achieved through a combination of tension and compression applied independently along two axes. Validation of simulations can be performed following these steps:

1. Evaluate the critical stress state in the simulation.

2. Manufacture a cruciform specimen using same material and layup as is in a critical area determined by simulation.

3. Induce the critical stress state in the specimen by biaxial test machine.

4. Compare failure stresses and strains obtained from both, the test and simulations.

The advantage of performing biaxial tests is that specimens are made from flat laminates, resulting in a relatively fast preparation compared to prototype experiments. The same instrumentation, measurement technique and evaluation method is used for all biaxial tests. On the other hand, in case of prototype testing, the prototype part has to be manufactured to be able perform experiment simulating conditions in real environment. In case of fiber composite part, this includes production of molds, which is usually the most expensive stage of manufacturing process. Additionally, specialized test rigs with comprehensive measurement instrumentation are required. The significant advantage of this approach is the robust validation of simulations, leading to trustworthy results. However, it comes at the expense of high costs.

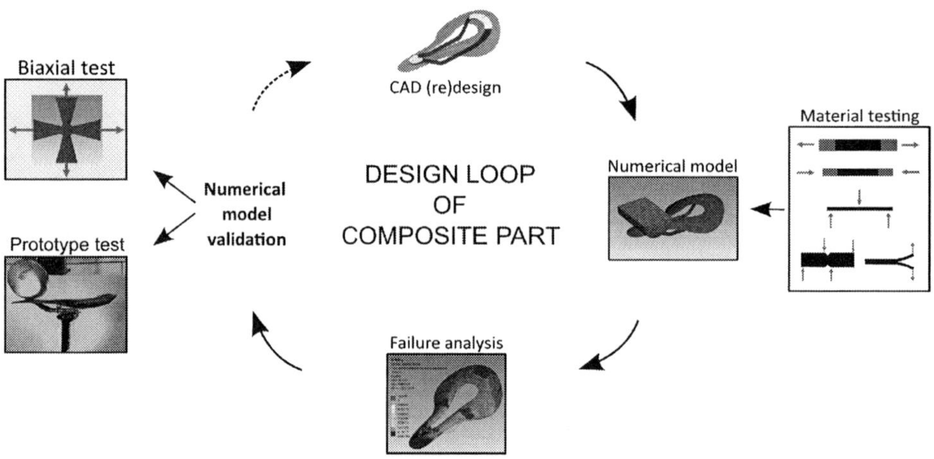

Fig. 1: Design loop of composite part.

Cruciform specimens must meet several criteria: i) failure should occur in the central region, ii) the area with a uniform biaxial stress state must be sufficiently large, iii) global shear stress in the central section should be minimized, and iv) the results must be repeatable [6]. A commonly used specimen type, as designed in [7], features a reduced thickness in the central region to promote failure at that location and includes double fillet corners at the junctions of the specimen arms.

The development of specimen type is still ongoing, as challenges remain with premature failures occurring outside the central section. Factors such as specimen geometry, tab types and materials, and the method of tab application (e.g., bonding or milling) can significantly influence the failure behavior. An ideal specimen should fully utilize the material's potential to provide accurate and reliable data for validation purposes.

2 Biaxial cruciform tests

Biaxial cruciform test machine was developed at VUTS, a.s. Test machine consists of 4 independent actuators with maximal load capacity of 10 kN. The stroke of the machine is 350 mm, which allows composite and elastomer testing. Tests can be performed in both displacement and load control mode. Displacements and strains are measured by Digital Image Correlation system Monet 3D. Detailed scheme of the biaxial testing setup is shown in Fig. 2.

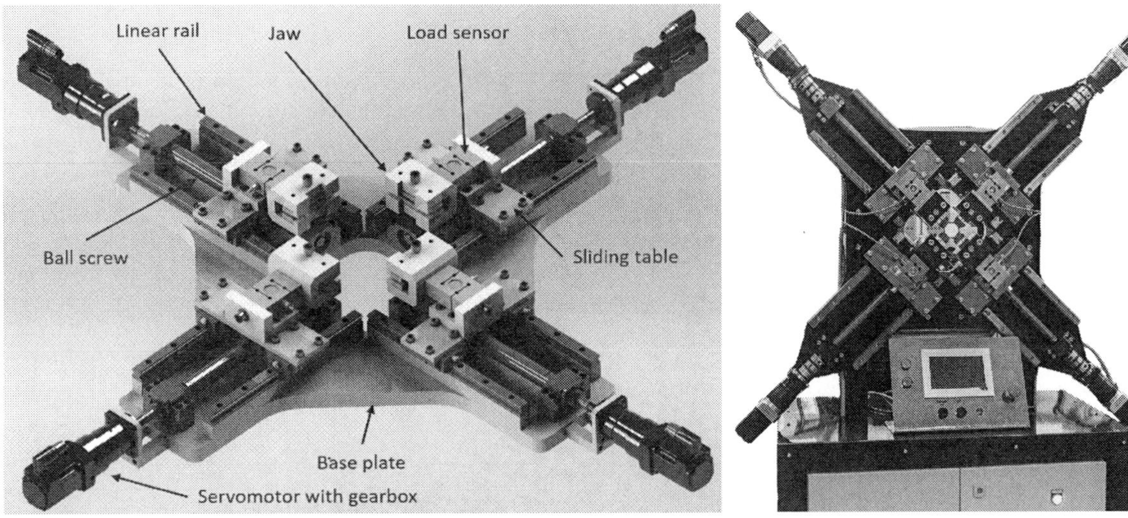

Fig. 2: Biaxial testing machine: CAD model scheme and real machine.

2.1 Analysed specimens

Equibiaxial tensile tests were performed in the displacement control mode. The cruciform specimen geometry features a double corner fillet with reduced thickness in the central area. The geometry arose from the geometry C developed in [7] and several adjustments were made (see Fig. 3). Original dimensions of this specimen are $0.7\times$ scaled down to be able to perform measurement on a test machine with maximal load 10 kN. The arms are not straight but towards the clamps they are wider to ensure good grip in the clamps. Laminates were produced by vacuum infusion process using LH385 epoxy system. Specimen were milled to final shapes by CNC machine and GF tabs were bonded to the CF laminate using Letoxit PL20 (see Fig. 3). Carbon-fiber laminate of interest featuring cross-ply layup $[0/90]_S$ and three different tab types are examined:

- no tabs,
- glass-fiber laminate tabs $[\pm 45F]_3$,
- glass-fiber laminate tabs $[0F]_3$.

Fig. 3: Specimen with GF tabs before bonding (left) and speckle pattern for DIC (right).

2.2 Experimental results

Strains are evaluated in central square 8x8 mm of specimen (see Fig. 3). Average values of measured failure strains, forces and computed strengths are summarized in Tab. 1. Strengths are computed using plane stress equations and mechanical properties noted in Tab. 2. The specimens set without tabs demonstrate consistent results with minimal measured value $\epsilon_f = 0.95$ % and maximal measured value $\epsilon_f = 1.15$ %. Both specimen sets with GF tabs show similar behavior reaching higher failure strains from $\epsilon_f = 1.27$ % to $\epsilon_f = 1.6$ % and strengths compared to the specimens with no tabs. The value of the highest reached failure strain $\epsilon_f = 1.6$ % is close to the value $\epsilon_{funiaxial} = 1.65$ % obtained by uniaxial tests on coupon specimens of same layup and material. The use of tabs significantly enhances the values of failure strain and strength.

Tab. 1: Measured average values of failure strain, computed strength and force at ultimate failure of different specimen tab types and number of tests performed per specimen type.

Experiment	ϵ_{fx} [%]	ϵ_{fy} [%]	X_t $[MPa]$	Y_t $[MPa]$	F_x $[N]$	F_y $[N]$	N
No tabs	1.08	1.06	651	640	5294	5263	10
GF $[45F]_3$	1.35	1.44	815	870	9089	9099	5
GF $[0F]_3$	1.37	1.51	831	908	9618	9523	3

3 Numerical model

A quarter numerical model was employed to simulate biaxial testing of cruciform specimens. The simulations were performed using the finite element software Ansys 2022R2. The boundary conditions were $u = 0.5$ mm in the end of the tabs, see Fig. 4. The computational mesh consists of 62000 elements of type SOLID185. The material model utilized in this study was orthotropic elasticity with a progressive damage approach. The progressive damage model incorporates the Puck failure criterion, which triggers a reduction in mechanical properties within the element when the failure criteria are met. The degradation factor, denoted as 1, corresponds to a 100 % reduction of stiffness, while a factor of 0 implies no reduction. Values of degradation factors were set to $E^*_{ft} = 0.99$ (Fiber tensile damage) and $E^*_{mt} = 0.85$ (Matrix tensile damage).

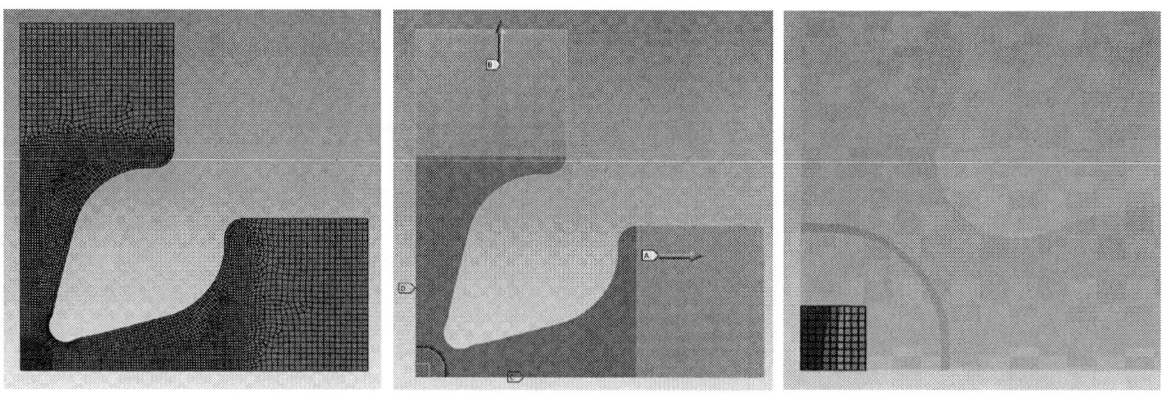

Fig. 4: Mesh, boundary conditions and observed area.

Mechanical properties (considered as mechanical properties at lamina level) of specimen and tabs are summarized in Tab. 2. Note that using linear elasticity model for glass-fiber laminate is simplification as its stress-strain curve is not linear, therefore secant tensile modulus is used.

3.1 Validation of numerical model

Firstly, simulations were performed using uniaxially measured value of tensile strength in fiber direction $X_t = 1620$ MPa. Force-strain curves from both simulation and experiment are compared in left section of

Tab. 2: Mechanical properties and stress limits of unidirectional CF laminate $[0]_4$ 125 gsm and woven GF laminate $[0]_4$ 200 gsm. E, G, X, Y and S in [MPa] and ν in [1].

	E_x	E_y	ν_{xy}	G_{xy}	X_t	Y_t	S_{xy}
CF UD	126 000	4700	0.365	4700	1620	29	60
GF fabric	13 800	13800	0.143	3300	392	392	120

Fig. 6. The experimental curves represent averaged data from the specimen set. Additionally, the figure illustrates the first ply failure in the matrix (marked with a black circle) and in the fibers (marked with a black square) obtained through the Puck failure criteria in simulations. The values of failure strains from experiments consistently exceed those from simulations. Tensile strength in fiber direction X_t determined by uniaxial test is the mechanical property of highest influence on failure strain ϵ_f.

Secondly, the force-strain curves from simulations were fitted by adjusting tensile strength in fiber direction X_t to align with experimental results. The force-strain curves exhibited the closest fit for tensile strength $X_{tfit} = 2087$ MPa, see right part of Fig. 6. The observed discrepancy underscores the need for a reassessment of the methodology employed in experimentally determining the tensile strength for application in the progressive damage model.

The comparison of strain maps obtained using experiment and simulations are depicted in Fig. 5.

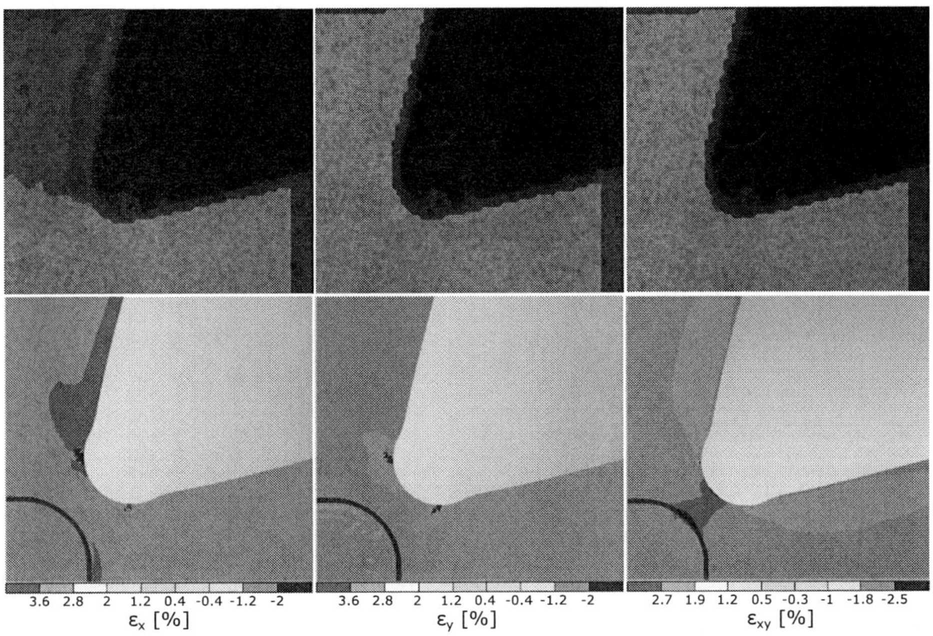

Fig. 5: Comparison of strain maps from experiment and simulation for specimen with GF $[45F]_3$ tabs at force 8000 N.

3.2 Observed failure

The failure observed in specimens, as depicted in Fig. 7, indicates successful biaxial tests for specimens with GF tabs. In contrast to specimens with no tabs, where failure occurs in a single arm.

However, the area of failure near transition between tabs and central section corresponds with maxima of Puck failure criteria observed in simulations. According to the simulations, the failure initiates in 0 fiber direction (see Fig. 7). It can be explained by stiffness change and therefore stress concentration in transition from tabs to central area, where the outer plies carrying load are more loaded than inner plies in second direction (which are protected by outer plies and load transition is smoother). This observation implies that the failures are still occurring prematurely.

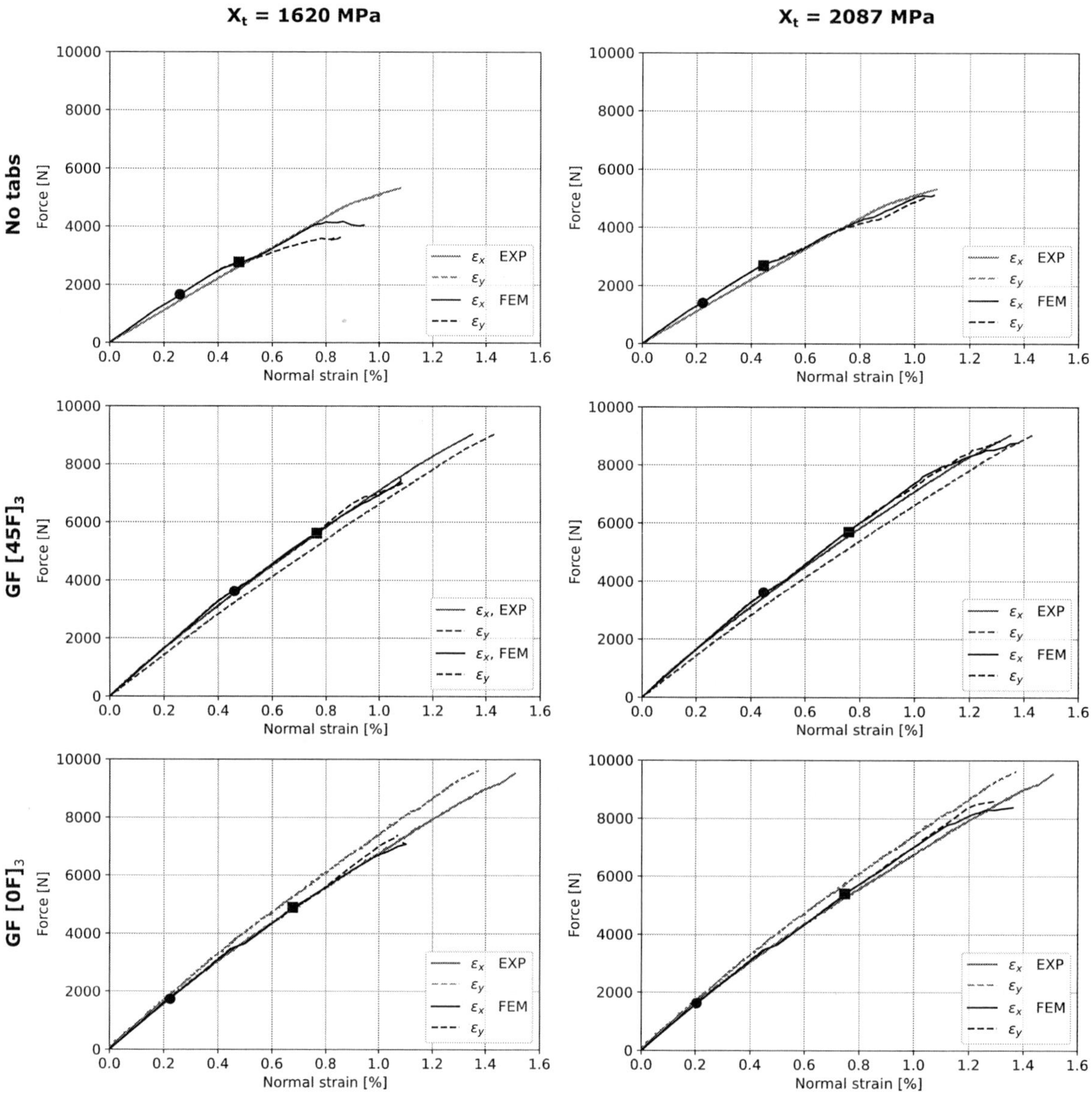

Fig. 6: Comparison of force-strain curves from simulation and averaged curves from experiment.

4 Conclusion

Biaxial cruciform tests can be used to validate failure criteria results obtained through numerical simulations, offering advantages over prototype testing in certain circumstances. A biaxial cruciform testing machine was developed and a testing methodology was adopted. Cruciform specimens were manufactured from carbon fiber - epoxy laminate without and with glass fiber tabs. Equibiaxial tension tests were performed, force and strain were measured and failure strains were evaluated.

Specimens with no tabs failed prematurely in arm section resulting in invalid biaxial test. In contrast, specimens with glass fiber tabs reached an average failure strain of approximately $\epsilon_f = 1.4$ %, still falling short of the uniaxially tested value $\epsilon_{f_{uniaxial}} = 1.65$ %. This indicates that the failure still might be premature, which is confirmed by numerical simulations with progressive damage model showing failure initiation in central section, but in the transition area between tabs and laminate of interest.

The comparison between experiments and simulations shows good agreement only in case when higher than uniaxially measured values of tensile strength were used as an input into progressive damage model. This

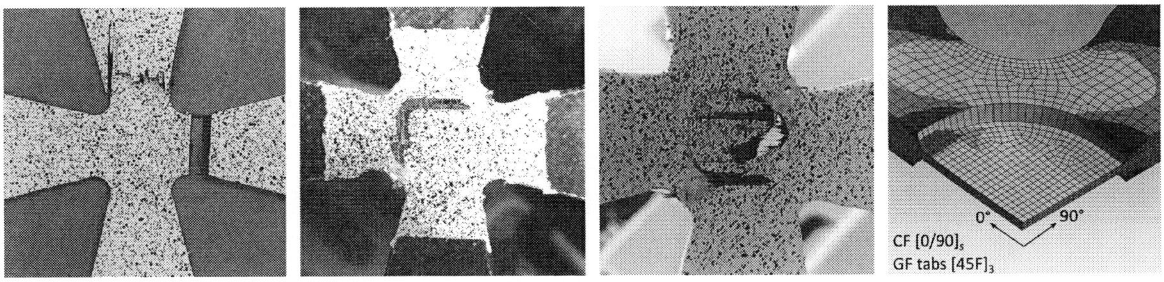

Fig. 7: Typical failure of specimens, from left 1) with no tabs 2) tabs GF$[45F]_3$, 3) tabs GF$[0F]_3$ and 4) maximum failure index near ultimate failure for tab types: GF $[45F]_3$

suggests that there may be a need to reconsider the method used to determine the uniaxial tensile strength for application in a progressive damage model.

The first ply failures for matrix and fibers were evaluated in simulations. The first ply failure strain values obtained in simulations are significantly lower than the ultimate failure strains measured experimentally, indicating that employing biaxial tests alongside simulations can be advantageous for determining the ultimate failure of components.

Acknowledgement

This publication was supported by the Ministry of Industry and Trade (MPO) within the framework of institutional support for long-term strategic development of the research organization - provider MPO, recipient VÚTS, a. s. This work was also supported by the Student Grant Scheme at the Technical University of Liberec through project nr. SGS-2023-3378.

References

[1] R. Zbončák, Pevnostní kritéria pro kompozitní materiály, Vydání první (Liberec: VÚTS, a.s., 2018), doi: 10.1617/s11527-013-0081-y.

[2] A. Smits et al., Design of a cruciform specimen for biaxial testing of fibre reinforced composite laminates, Composites Science and Technology 66, no. 7-8, 964–975 (2006), doi: 10.1016/j.compscitech.2005.08.011.

[3] S. Horta Muñoz and M. del C. Serna Moreno, Advances in cruciform biaxial testing of fibre-reinforced polymers, Polymers 14, no. 4, 686 (2022), doi: 10.3390/polym14040686.

[4] C. Ramault, Guidelines for biaxial testing of fibre reinforced composites using a cruciform specimen, Ph.D. thesis, Vrije Universiteit Brussel, 2012.

[5] P.D. Soden, M.J. Hinton, and A.S. Kaddour, Biaxial test results for strength and deformation of a range of E-glass and carbon fibre reinforced composite laminates: failure exercise benchmark data, Composites Science and Technology 62, no. 12-13, 1489–1514 (2002), doi: 10.1016/S0266-3538(02)00093-3.

[6] J.P. Boehler, S. Demmerle, and S. Koss, A new direct biaxial testing machine for anisotropic materials, doi: 10.1007/BF02328435.

[7] A. Smits, D. Van Hemelrijck, T.P. Philippidis, and A. Cardon, Design of a cruciform specimen for biaxial testing of fibre reinforced composite laminates, Composites Science and Technology 66, no. 7-8, 964–975 (2006), doi: 10.1016/j.compscitech.2005.08.011.

Efficiency of Reinforcement Materials on Bus Frame Stability

V. Yanovych [1,2,a], J. Václavík [1]

[1] *Research and Testing Institute Plzen, Dynamic Testing Laboratory, Czech Republic*
[2] *Department of Power System Engineering, Faculty of Mechanical Engineering, University of West Bohemia, Czech Republic*
[a] *yanovych@vzuplzen.cz*

Abstract: This paper presents the results of an experimental investigation of the CFRP and epoxy foam's effect on bus frame stability. As part of the research, a series of quasi-static tests were conducted to analyze the behavior of the test specimen in four experimental scenarios. Initially, the steel structure was tested, followed by reinforcement with carbon fiber on one post, then extending the carbon fiber reinforcement to both the crossbar and side posts. The final scenario involved filling the steel structure with epoxy foam, excluding carbon fiber. Generally, according to the results obtained, the reinforcing materials have a significant impact on the structural stability of the specimen structure. Namely, the deformation energy in the case of foam and CFRP reinforcement decreased by about 15 %.

Keywords: bus frame; reinforcement materials; structural stability; deformation.

1 Introduction

Bus accidents occur frequently worldwide. In Europe, approximately 150K individuals sustain injuries, while 150 lose their lives annually due to such incidents [1]. In the United States, according to the National Highway Traffic Safety Administration's report, there were 16K injuries and fatalities resulting from bus crashes, with over half of the fatalities arising from non-collision accidents [2]. Consequently, prioritizing the safety of bus passengers has become a critical issue in the bus industry, reflecting the escalating safety challenges seen in the automotive sector. Safeguarding the physical safety of bus passengers isn't solely a matter of marketing; it's also a duty aligned with international standards.

Currently, one of the most promising methods of increasing the structural stability of passenger vehicles is the application of carbon fiber-reinforced polymer (CFRP) and epoxy foam [3-5]. A large number of different studies have been devoted to this issue. However, the reinforcement process is usually adapted to the specific characteristics of each structure, resulting in a customized investigation of each experimental case.

2 Test Configuration and Measuring Equipment

Experimental investigations were conducted at the Dynamic Testing Laboratory of the Research and Testing Institute in Pilsen s.r.o. (Czech Republic). The primary focus of the study was to assess the effectiveness of various reinforcing materials on the structural stability of the frame in passenger transport, specifically in the event of a bus rollover accident. In such situations, the roof and sidewalls undergo significant deformation, which can lead to a critical reduction in safe space, resulting in injury to passengers.

In the current study, a standard bus frame section was used as the test specimen. The experimental plan included conducting a series of quasi-static tests to analyze the behavior of the test specimen across four experimental scenarios. Initially, the investigation focused on a structure comprising solely of steel. After that, in the second scenario, the frame received reinforcement by carbon fiber exclusively on one side. Moving on to the third case, carbon fiber reinforcement (CFRP) was extended to cover both posts and the crossbar of the frame. It is crucial to note that in this configuration, the upper reinforcing lamella was positioned between welds. Finally, the fourth scenario involved filling the steel structure with epoxy foam, without adding carbon fiber.

Figure 1 illustrates the organization of the test bench for assessing the structural stability of the specimen under various scenarios. Classically, a hydraulic cylinder has been applied to simulate the force load on the frame, modeling conditions similar to those that occur when a bus rolls over. One end of the cylinder was attached to a separate base positioned at half the height of the frame, while the other end was linked to the upper corner of the segment via a load cell HBM S9M 20kN. Consequently, when the hydraulic cylinder exerted force, the load cell recorded the force characteristics at which the specimen deformed. It should be noted that in all experimental scenarios, a force of 10 kN was applied at 22 degrees relative to the horizontal plane.

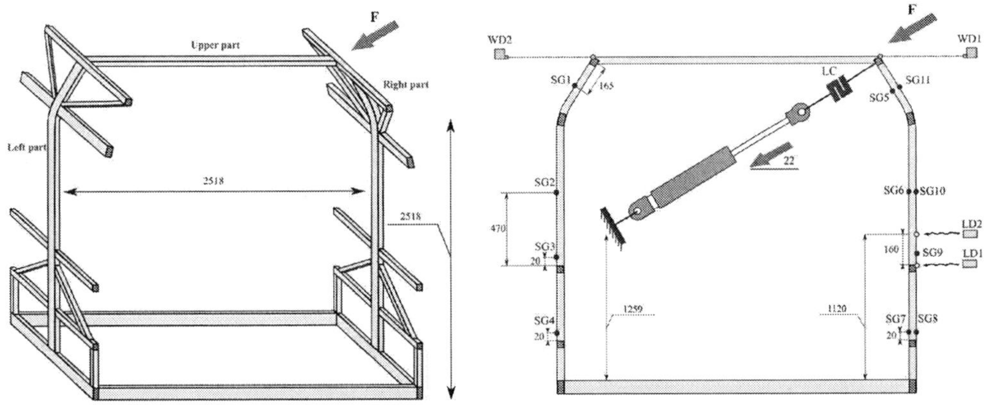

Fig. 1: Experimental setup, there a test specimen sketch on the left and a test bench on the right. Where SG - strain gauges, WD - wire digital sensors WDS-1000-P60, LD - laser distance sensors HG-C1200, LC - load cell HBM S9M/20kN.

Fig. 2: Practical implementation of the test bench and the placement of measuring sensors.

To quantify the deformation pattern of the specimen, a variety of measurement sensors were employed. Figure 2 depict the practical implementation of the test bench and the placement of measuring sensors. To assess the displacement of the specimen construction under load, two digital wire sensors WDS-1000-P60, were positioned at the upper corners of the frame. Moreover, two laser distance sensors, namely the HG-C1200, were strategically placed in the window area of the bus, where the gap between the frame and passengers is minimal. Therefore, in the event of extraordinary situations such as bus rollovers, the reduction of this space becomes critical for ensuring passenger safety. Furthermore, eleven strain gauges were used on both frame supports to evaluate the effect of different reinforcing materials on the structural stability of the specimen. Specifically, on the post where the deformation force was applied, seven sensors were positioned on both the inner and outer sides. Conversely, on the opposite spots, only four strain gauges were situated, exclusively on the outside. Experimental data were gathered and analyzed using the Catman software environment, with data collected at a frequency of 600 Hz from all the aforementioned devices.

3 Experimental Results

The experimental results in Fig. 3 show that in all test scenarios, the external load causes clear localized deformation zones that resemble bulges at the joints of the upper and side frame elements. It is notable that the orientation of the bulges differs depending on the side of the specimen. Specifically, the bulge on the right side displays a downward curvature, whereas on the left side, it curves upward. This contrast is attributed to the compression applied by the hydraulic cylinder on the right side. As a result, the greatest stress and subsequent plastic deformation of the structure are observed in this area. However, depending on the type of specimen reinforcement, the resulting bulges differ.

In the scenario involving the unreinforced frame (see Fig. 3), bulging occurs at distances of 22 mm and 41 mm from the right and left posts, respectively. However, in the case of carbon fiber reinforcement solely on the left side (see Fig. 4), the bulge shifts away from the right posts by 55 mm while remaining 41 mm from the left one. Whereas, in the foam reinforced scenario, the bulges appear symmetrically on both specimen sides at the mounting hole's location at 51-55 mm (see Fig. 5). Moreover, the local concentration of significant stresses in this area leads to the appearance of cracks.

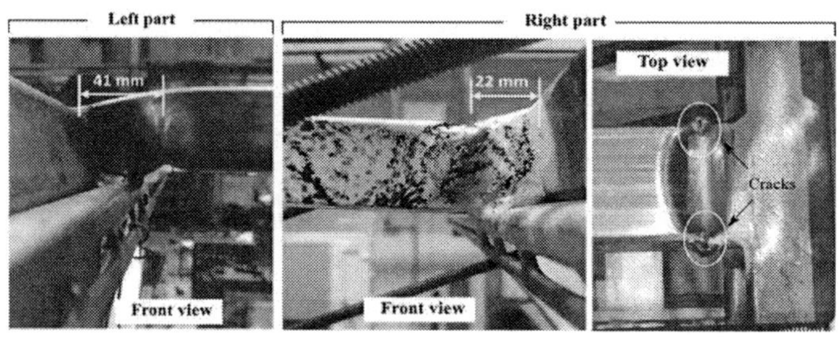

Fig. 3: Local deformation of the unreinforced specimen.

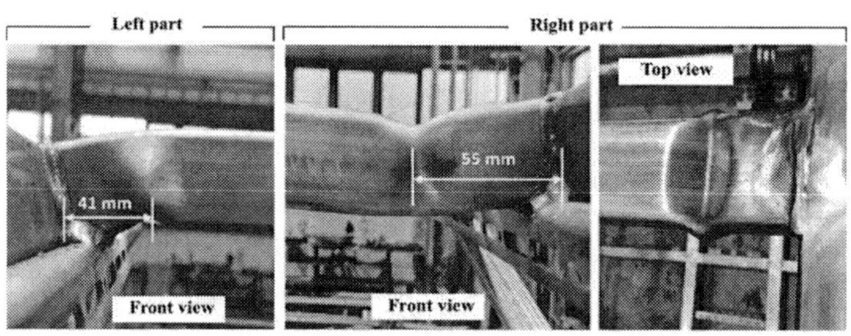

Fig. 4: Local deformation of the reinforced specimen with carbon fiber on one side.

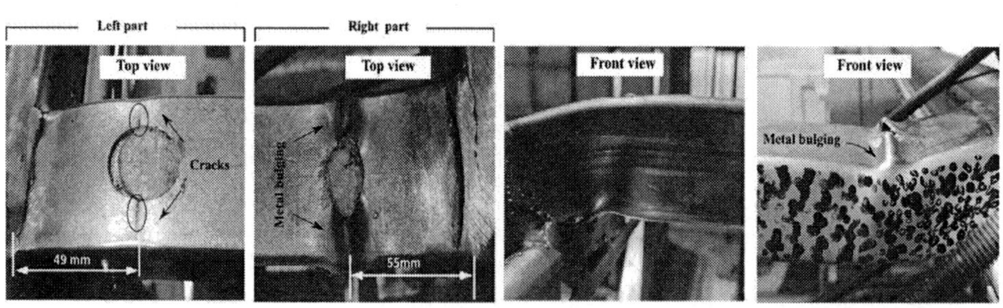

Fig. 5: Local deformation of the foam reinforced specimen.

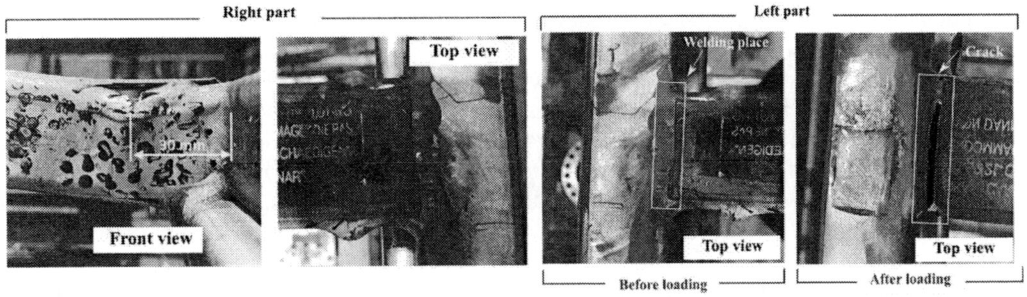

Fig. 6: Local deformation of the specimen reinforced with carbon fiber on both the crossbar and side posts.

The most pronounced impact of reinforcement becomes evident in the scenario involving the carbon fiber reinforcement of the crossbar and side posts. In particular, as a result of the effective stress redistribution in the specimen structure, a welded joint between the crossbar and the left post ruptured (see Fig. 6). It should be noted that this phenomenon was not observed in other test cases. Thus, in this scenario, the bulge is observed only on the right side, at a distance of 30 mm from the post.

As depicted in Fig. 7, the maximum displacement of the specimen construction across all experimental scenarios reached approximately 600 mm. This trend occurs because the applied deforming load of 10 kN exceeds the frame's ability to withstand deformation. Additionally, variations in maximum displacement between experimental cases are explained by tightening or loosening the connection between the hydraulic cylinder and load cell. The plateau observed on the graph indicates the stage of complete compression.

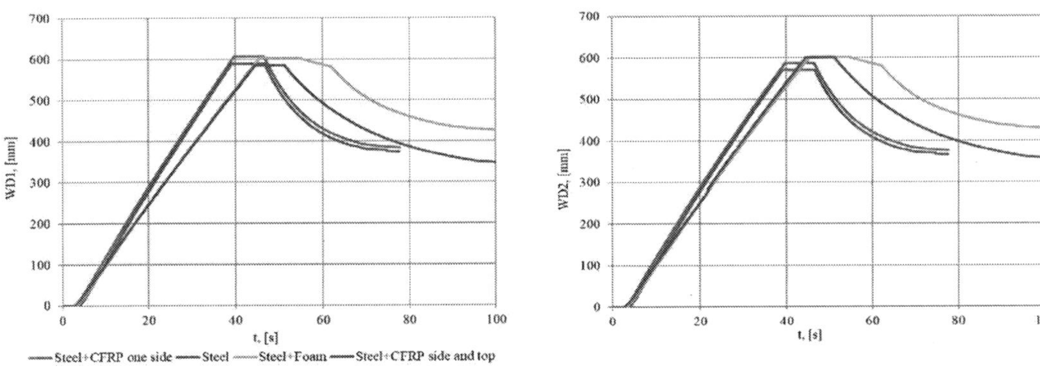

Fig. 7: Specimen displacement across different experimental scenarios. The left and right graphs correspond to the sides of the frame.

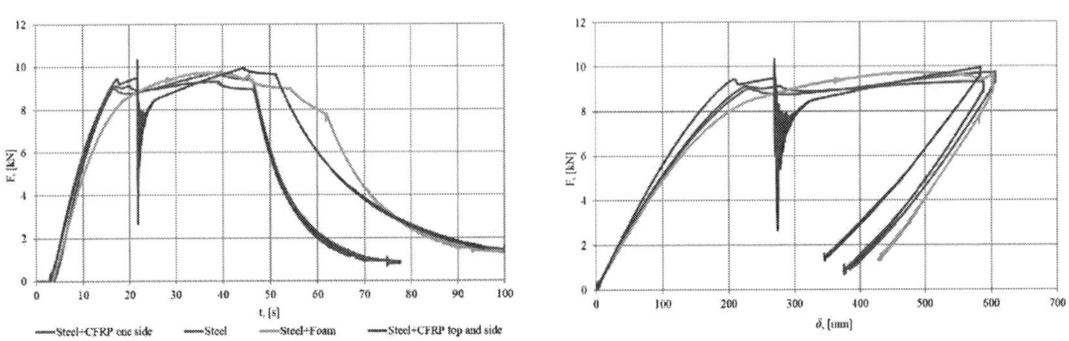

Fig. 8: Force-deformation dependence for all experimental scenarios.

Figure 8 depicts the force-displacement relationship for all experimental scenarios. Notably, the maximum deformation force is evident for specimen reinforced with carbon fiber on both the crossbar and side posts. Additionally, the obtained results exhibit a significant data perturbation at 22 seconds, which is attributed to the weld failure. It's worth noting that for all experimental cases, except for epoxy foam, the deformation curves exhibit clearly defined yield points. That indicates the elastic behavior of the specimen construction in this case. In contrast, in the foam scenario, a more pronounced plastic behavior is observed.

Additionally, the elastic deformation energy of the specimen was estimated based on the obtained data, considering the type of reinforcement. Calculation data in Fig. 3 showed that in the cases of epoxy foam and carbon fiber reinforcement on the crossbar and side posts, the deformation energy decreased by approximately 15%. Consequently, under these conditions, the frame requires less energy to maintain its shape under external loads. This achieved due to the higher strength of the structure, which efficiently absorbs and distributes energy during deformation, thus reducing energy losses attributed to elastic processes.

However, the data in Fig. 10 indicate that the carbon fiber reinforcement case is also characterized by the largest displacement of the structure. This contradiction is explained by the peculiarities of the stress distribution within the specimen construction. In a typical case, for an unreinforced specimen, a substantial portion of the external load energy rapidly converts into plastic deformation of the upper part of the frame. Meanwhile, the residual energy results in only a small displacement of the construction.

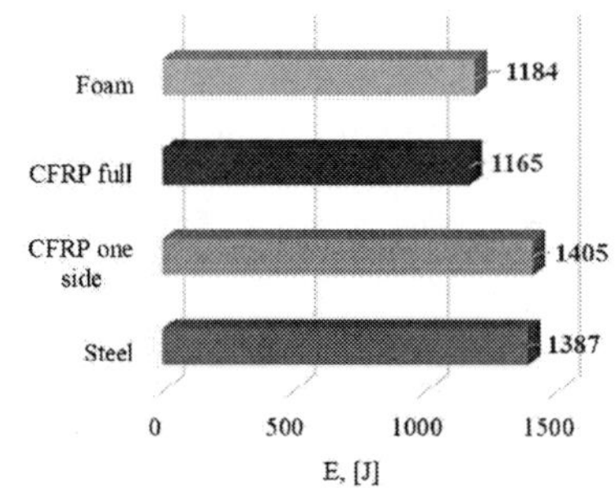

Fig. 9: Elastic deformation energy depending on the bus frame reinforcement type.

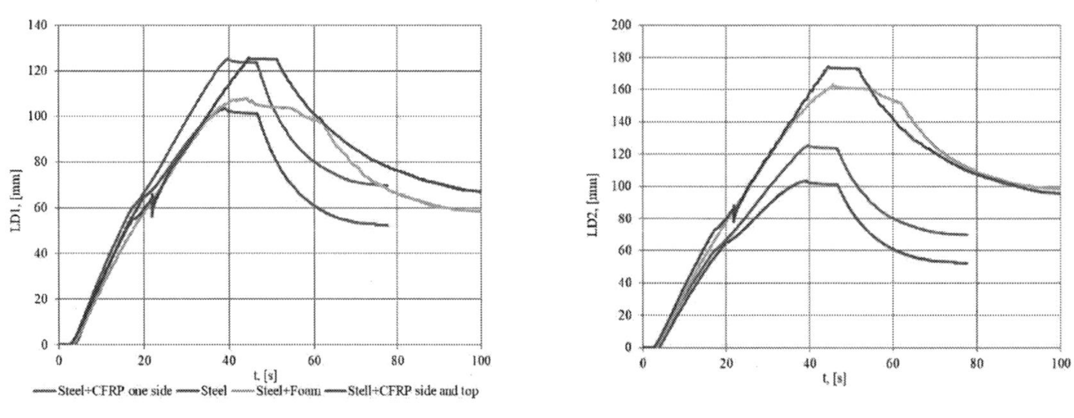

Fig. 10: Specimen displacement across different experimental scenarios.

In general, reinforcing materials have a significant impact on the structural stability of the sample structure. This effect is particularly evident in Fig. 11, which shows the strain estimation at measurement point 9. It can be observed that the foam and carbon fiber reinforcement led to approximately a 65% and 48% reduction in strain, respectively. Unfortunately, as already mentioned, in the case of carbon fiber on the crossbar and side posts, the data is only reliable up to 20 seconds due to structural weld failure.

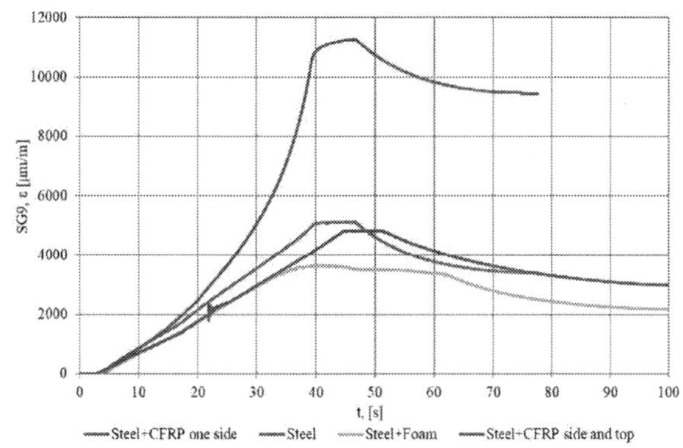

Fig. 11: Strain-time distribution at measurement point 9 depending on type reinforcement.

4 Conclusion

In the classical case of a frame without reinforcement, the structure undergoes significant plastic deformations due to the applied load. In contrast, the epoxy foam acts as an absorber of the deformation energy and helps to effectively distribute the resulting stresses evenly throughout the frame structure. Therefore, the plastic deformation on both sides of the frame is similar. This statement is also confirmed by the configuration of the resulting bulges, which are much wider than in other reinforcement test cases. At the same time, carbon fiber reinforcement increases the stiffness of the frame by effectively redistributing the stress state of the structure. However, it is important to emphasize that, unlike epoxy foam, carbon fiber does not absorb energy, but only redistributes it. This distinct feature causes a localized concentration of significant stress at the welded joints of the frame. Consequently, this concentration leads to the failure of the welded joint outside the carbon fiber reinforcement. Generally, it can be concluded that carbon fiber and foam reinforcement are effective approach to improving the stability of bus frame structures. Namely, their application reduced the deformation energy by almost 15%.

Acknowledgement

The work was performed within the framework of the M-ERA.NET 2019 competition and was supported by the Technology Agency of the Czech Republic under the number TH71020003.

References

[1] P. Albertsson, T. Falkmer, Is there a pattern in European bus and coach incidents? A literature analysis with special focus on injury causation and injury mechanisms, Accident Analysis and Prevention 37(2) (2005) 225–233, https://doi.org/10.1016/j.aap.2004.03.006.

[2] K. Kyoung-Tak et al., Design of a composite roll bar for the improvement of bus rollover crashworthiness, Composites Part B: Engineering 43 (2012) 1705–1713, http://dx.doi.org/10.1016/j.compositesb.2012.01.072.

[3] A. Pawlak et al., The Use of CFRP for Structural Reinforcement-Literature Review, Metals 12 (2022) 1470, http://dx.doi.org/10.3390/met12091470.

[4] R. Sen, L. Liby, A. Mullins, Strengthening steel bridge sections using CFRP laminates, Composites Part B: Engineering 32 (2001) 309–322, http://dx.doi.org/10.1016/S1359-8368(01)00006-3.

[5] J.G. Teng et al., Behaviour and strength of FRP-strengthened RC structures: A state-of-the-art review, Proceedings of the Institution of Civil Engineers-Structures and Buildings 156 (2003) 334–335, http://dx.doi.org/10.1680/stbu.156.3.334.37862.